10653286

Mechanical Measurements

Mechanical Measurements

R. S. Sirohi
and
H. C. Radha Krishna
Department of Mechanical Engineering
Indian Institute of Technology, Madras

A HALSTED PRESS BOOK

JOHN WILEY & SONS
New York London Sydney Toronto

Published in the U.S.A., Canada and
Latin America by Halsted Press,
a Division of John Wiley & Sons, Inc., New York

Library of Congress Cataloging in Publication Data

Sirohi, R S
Mechanical measurements.

"A Halsted Press book."
Bibliography: p.
Includes index.
1. Mensuration. 2. Physical measurements.

I. Radha Krishna, H. C., joint author. II. Title.
T50.S58 620'.0044 80-27233
ISBN 0-470-27107-8

Printed in India at Urvashi Press, Meerut.

Foreword

Experimental measurements, methods and techniques are becoming increasingly important in engineering education. The ability to perform experimentation and take measurements with acceptable precision deserve more emphasis. In recent years laboratory programmes have been modernized, sophisticated electronic instrumentation are incorporated into the programme and newer techniques have been developed. In as much as a well designed laboratory programme is essential in the undergraduate engineering education, providing the students with laboratory manuals, guides and text books is equally important. It is in this context that this book makes its timely appearance, which, I am sure, will be gladly received by all. My senior colleagues, Dr. R. S. Sirohi and Dr. H. C. Radha Krishna, are to be congratulated for their efforts in bringing out this book to meet the needs of the students and teachers in comprehending the various aspects of mechanical measurements.

Indian Institute of Technology
Madras

Dr. M.C. Gupta
Professor of Mechanical Engineering
and
Coordinator Mechanical Engineering
Education Development Centre

Preface

Engineers are involved in design of systems, systems involving many components. The effectiveness of the system is verified by Measurements. Measurement is an Art, a Science and a Technique by itself. There is nothing which can be effected without Measurement. An attempt is made here to deal with a part of the knowledge commonly identified as Mechanical Measurements. The stress is laid on the understanding of the physics of the Measurement Techniques.

Every measurement is in error. For the measurement to be meaningful, the nature and magnitude of the error should be known. Therefore, the book begins with error analysis and application of statistical principles to measurements.

The methods of measuring various mechanical quantities are discussed subsequently, covering both the basic and derived quantities. Effort has been to present the subject in S.I. Units. The coverage in the book is such that it may be successfully used as a text both for graduate and post-graduate classes and a constant reference by a researcher.

If the student realises the importance and understands the method of measurement required by him, we feel highly pleased and honoured. The material contained in the book is extracted freely from books, journals and pamphlets. We gratefully acknowledge all of them. Professor M. C. Gupta, Professor of Mechanical Engineering and Coordinator, Curriculum Development Centre, Indian Institute of Technology, Madras spontaneously responded to our request to publish this first in the form of a monograph. We are very thankful to him for his encouragement and support. Many of our colleagues have been of immense help in this effort of ours. We acknowledge all of them.

R. S. Sirohi
H. C. Radha Krishna

Contents

1 Introduction

1.1 What is measurement?

Fundamentally, measurement is the act or the result of a quantitative comparison between a predefined standard and an unknown magnitude. If the result is to be meaningful, two requirements must be met in the act of measurement:

(i) the standard must be accurately known and internationally accepted, and

(ii) the apparatus and procedure employed for obtaining the comparison must be provable.

In order to be able to consistently compare quantitatively, certain standards of length, mass, time, temperature and electrical quantities* have been established. These standards are internationally accepted and well preserved under controlled environmental conditions.

The standard of length is the standard metre defined as a length between two very fine lines engraved on a platinum-iridium bar maintained and measured under very accurate conditions. The General Conference on Weights and Measures defined the standard metre in terms of wavelength of the orange-red light of Kr^{86} lamp. The standard metre is thus equivalent to 1,650,763.73 wavelengths of Kr^{86} orange-red light. The kilogram is defined in terms of platinum-iridium mass. Both these standards are maintained at International Bureau of Weights and Measures in Sevres, France. Cesium clock has been accepted as the standard of time measurement.

An absolute temperature scale was conceived by Lord Kelvin in 1854 on the basis of second law of thermodynamics. The international temperature scale of 1948 furnishes an experimental basis of a temperature scale which approximates as closely as possible with absolute thermodynamic scale.

*The standard units of electrical quantities are derivable from the mechanical units of force, mass, length and time.

1.2 Fundamental methods of measurements

There are two basic methods of measurement:

(i) direct comparison with the primary or secondary standard, and
(ii) indirect comparison with a standard, through a calibrated system.

Both these methods are employed depending on the requirement. But to save the primary standard from a frequent and direct handling, secondary standards are generally used for direct comparison or calibration.

1.3 Calibration

The calibration of all instruments is important, for it affords the opportunity to check the instrument against a known standard and subsequently reduce the error in the measurement. Calibration procedures involve a comparison of a particular instrument with either (a) a primary standard, (b) a secondary standard with higher accuracy than the instrument to be calibrated, or (c) a known input source. The calibration procedure for various instruments will be discussed in the later chapters.

1.4 Why to make measurements?

A very basic function of all engineering branches is design—design of systems consisting of several elements which are expected to function in a particular fashion. The measurement is required to test the functioning of components which constitute the system and finally the function of the system itself.

Further, in basic sciences fundamental principles or derived phenomena are studied in great detail to confirm the validity of certain postulates, or sometimes in the anticipation of discovering some strange behaviour of nature, a sophisticated instrumentation is required. The analysis is carried out on the numerical results which arise due to the act of measurement.

To propound a statistical law, the measurements are made on a number of systems.

Hence measurements are essential for evaluating the performance of a system, or studying its response to a particular input function, or studying some basic law of nature, etc. The measuring instrument is an essential component of an automatic control system.

1.5 Concept of a generalised measurement system

An instrument is designed to perform a certain task and its description is, therefore, always possible in terms of its physical elements. However, this approach has its shortcomings—particularly it demands a separate description for each instrument. The approach can be generalised if the instrument is described in terms of its functional element. It will be found that a physical element may perform many functions. There could be various schemes to describe the instrument in terms of its functional elements, but the one discussed below is found to be quite adequate and simple.

A block diagram representation of a generalised measurement instrumentation is shown in Fig. 1.1.

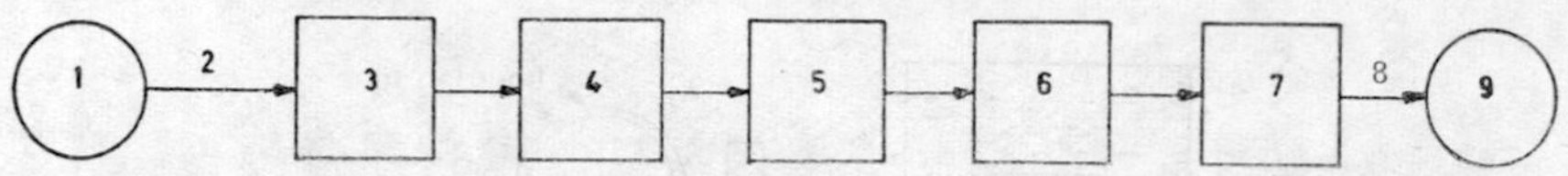

Fig. 1.1 Block diagram of a generalised measurement instrumentation
1. Measured medium; 2. Measured quantity; 3. Primary sensing element; 4. Variable conversion element; 5. Variable manipulation element; 6. Data transmission element; 7. Data presentation element; 8. Presented data; 9. Observer.

(a) PRIMARY SENSING ELEMENT

It is an element which first receives energy from the medium to be measured and produces a proportional output. The output signal of the primary sensing element is a physical variable such as displacement or voltage. The primary sensing element, therefore, is a transducer which converts (transduces) one physical variable or effect into another. For the measurement to be faithful, the transducer should be so designed as to extract a very small amount of energy from the medium. In other words, the medium should not be disturbed appreciably by inserting the sensing element.

(b) VARIABLE-CONVERSION ELEMENT

The output signal of the primary sensing element may be required to be converted to a more suitable variable while preserving its information content. This function is performed by variable-conversion element and may be considered as the second transducer stage.

(c) VARIABLE-MANIPULATION ELEMENT

This element is an intermediate stage of a measuring system. This modifies the direct signal by amplification, filtering or other means so that a desired output is produced. During this stage, the physical nature of the variable remains unaltered.

(d) DATA-TRANSMISSION ELEMENT

When the functional elements of an instrument are spatially separated, it becomes necessary to transmit signal from one element to another. This function is performed by data transmission element.

(e) DATA PRESENTATION ELEMENT

Usually the information about the quantity to be measured is to be communicated to a human being for monitoring, control or analysis purposes. This is, therefore, to be presented in a form recognisable by some human senses. If the information is to be presented to the computer it can be done in the form of binary scale on the punched tape or cards. An element that performs this 'translation' function is called a data presentation element.

An example given below may illustrate the working of this representation. A pressure-type thermometer (Fig. 1.2(a)) is used to measure the temperature of a fluid. The thermometer works on the principle of

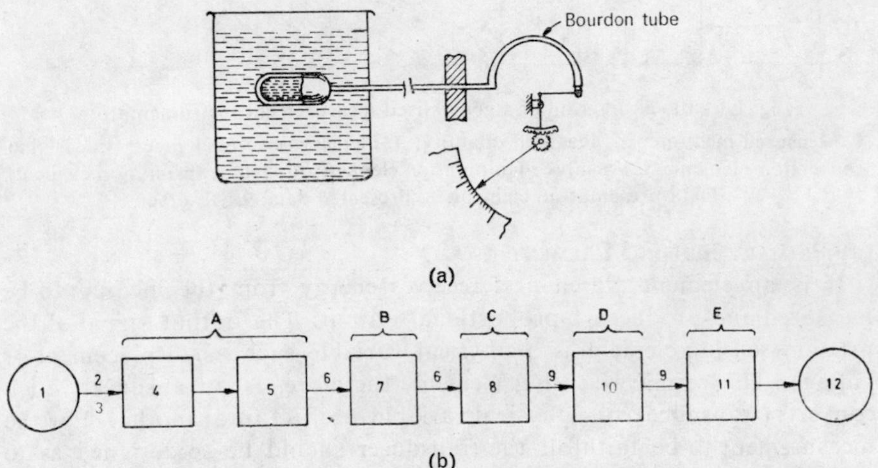

Fig. 1.2 Pressure thermometer

1. Measured medium; 2. Temperature; 3. Measured quantity; 4. Primary sensing element; 5. Variable conversion element; 6. Pressure; 7. Data transmission element; 8. Variable conversion element; 9. Motion; 10. Variable manipulation element; 11. Data presentation element; 12. Observer. A—Bulb; B—Tubing; C—Bourdon tube; D—Linkage and gear; E—Scale and pointer.

differential expansion of liquid, which in turn imparts pressure to the Bourdon tube. Through a rack and pinion arrangement, the deformation (displacement) of the Bourdon tube is magnified and read on the scale. Fig. 1.2(b) is a block-diagram representation of the measurement act. It may be emphasized that the same physical element can perform more than one function and the functions need not be performed in the sequence of Fig. 1.1.

When an overall description of an instrument is desired, the instrument may be considered to be doing an operation on the input quantity and producing an output. A block diagram of this is illustrated in Fig. 1.3.

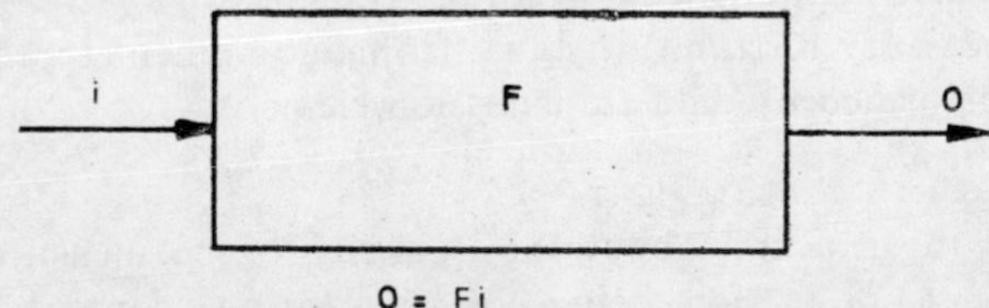

Fig. 1.3 Input-output relation of a measurement system

The input-output relationship is characterised by the operation F such that

$$o = Fi$$

where o is the output and i is the input.

The input to the instrument may be any one or combination of the following types:

(i) *Desired input* represents a quantity which the instrument is specifically meant to measure.

(ii) *Interfering input* represents a quantity to which the instrument is unintentionally sensitive.

(iii) *Modifying input* represents a quantity which modifies the input-output relationship for the desired and interfering inputs. The integrated influence of these inputs on the instrument's output is shown in Fig. 1.4.

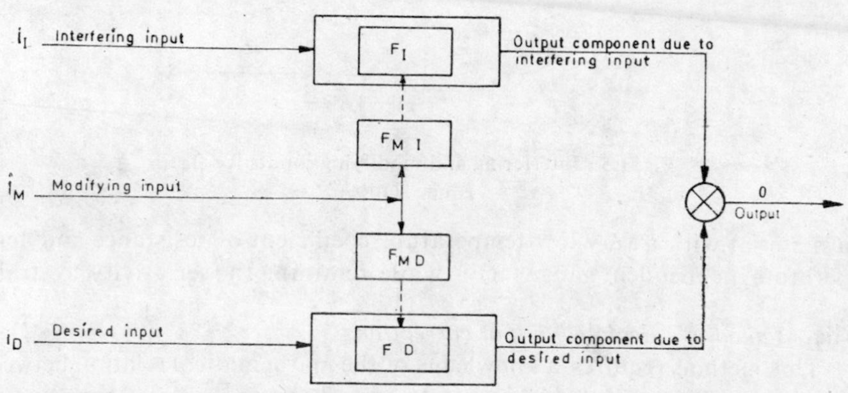

Fig. 1.4 Generalised input-output configuration

The physical meaning of these inputs can be grasped by considering an example from the measurement field. We consider an electrical strain gauge setup (Fig. 1.5) used for the measurement of strain induced due to the application of load. The strain ϵ is measured in terms of resistance change ΔR of the strain-gauge given by

$$\Delta R = G_F R \epsilon$$

where R is the resistance of the gauge and G_F the gauge factor. The resistance change is measured with Wheatstone bridge arrangement. In this, strain ϵ is the desired input and the bridge output e_0 is the output, which is proportional to ΔR.

One kind of interfering input to this instrument is a 50 Hz stray field which can induce voltage in the output although the strain is zero. The other interfering input is temperature. Any change in the ambient temperature results in the resistance change and hence an output is developed even in the absence of strain. Further due to the dependence of gauge factor G_F on temperature the proportionality constant is also affected by the ambient temperature change. Therefore, temperature is a modifying input as well.

Often it is desired to reduce or eliminate the influence of undesired inputs on the instrument. There are various ways to achieve this. A few are described below:

(i) *Method of inherent insensitivity*

The design of the instrument should be such that it is sensitive only to the desired input. In the example of Fig. 1.5, the strain gauge is made of

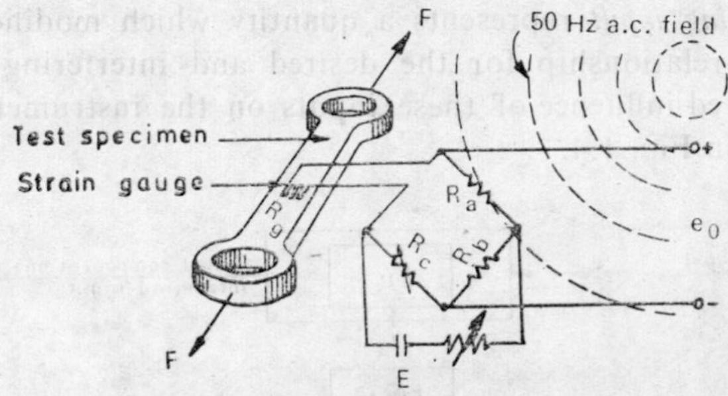

Fig. 1.5 Interfering and modifying inputs for strain gauge setup

a material with a very low temperature coefficient of resistance and temperature independent gauge factor while retaining the sensitivity to strain.

(ii) *Method of calculated output correction*

This method requires a knowledge of the mathematical relation between the spurious input and its output. In the above example, if temperature coefficient of resistance, temperature sensitivity of gauge and its temperature are known, then the correction to the output can be found and applied.

(iii) *Method of signal filtering*

This method is based on the possibility of introducing certain elements (filter) into the instrument which block the spurious inputs so that their effects on the output are minimised or eliminated. The filter can be inserted either in the path of spurious inputs or output path. As an example in the case of strain gauge, the spurious input due to 50 Hz stray field can be removed by shielding the instrument. The instruments sensitive to vibrations can be mounted on damping devices; the effect of tilt can be removed by gyroscopic mounting.

(iv) *Method of opposing inputs*

This method consists of intentionally introducing into the instrument interfering and/or modifying inputs that tend to cancel the bad effects of the unavoidable spurious inputs. Again in the case of strain gauge, the influence of temperature is removed by inserting one more identical strain gauge into another arm of the Wheatstone bridge and exposing it to the identical environmental conditions. There are many instruments which employ these compensation techniques, viz., resistance and thermocouple thermometry, photomultiplier circuitry, etc.

(v) *Analog signals*

The analog signals vary in continuous fashion. They can take infinite values in any given range. The devices which produce such signals are called analog devices. Most of the devices or instruments which are used for measurement and control are of analog type.

(vi) *Digital signals*

The digital signals vary in discrete steps and thus can take up only a finite number of different values in a given range. The devices or instruments that produce such signals are called digital devices. Due to the application of digital computers for data handling, reduction, and in automatic control, the importance of digital instrumentation is increasing very fast. The data to a digital computer is to be provided in the digital form, while most of the measuring instruments are of analog form, a conversion from analog to digital is required. This is performed by analog to digital converter. Similarly, to convert a digital signal to analog signal, digital-to-analog converter is needed. These devices serve as 'translators' that enable the computer to communicate with the outside world which is largely of analog nature.

(vii) *Deflection and null methods of operation*

The measuring instruments are used either in deflection or null mode. In deflection mode, the quantity to be measured produces some physical effect on a part that causes a similar but opposing effect on some other part of the instrument. At balance, the opposing effect is equal to that produced by the quantity to be measured. An example of such instruments is a moving coil galvanometer, where the force due to the current flowing in the coil is counterbalanced by the torsional force of the suspension filament. At balance the deflection of the light beam or the pointer gives the magnitude of the current. Such instruments are calibrated initially. In contrast to the deflection type instruments, a null type instrument maintains balance at one point between the effect generated by the quantity to be measured and the suitable opposing effect applied to it. A detector of imbalance and a means of restoring balance are the essential elements for such instruments. Knowledge of the quantity that produces opposing effect provides the value of the quantity to be measured.

Exercises

1. Define the word transducer. What do you understand by active and passive transducers? Give examples and explain their relative merits and demerits.

2. The data to the observer can be presented either in analog or digital form. Mention some methods for analog to digital conversion.
3. Null method of measurement is often preferred over the deflection method. Comment.
4. The natural frequency of oscillation of the balance wheel in a watch depends on the moment of inertia of the wheel and the spring constant of the (torsional) hair spring. A temperature rise results in a reduced spring constant, lowering the natural frequency. Propose a compensating means for this effect. Non-temperature sensitive material for hair spring is not an acceptable solution.
5. The pressure thermometer of Fig. 1.2(a) is used to measure the temperature of a remotely located hot body. The variation of ambient temperature will influence the temperature of fluid in the data transmission element, i.e., tubing and hence the measured temperature. How do you compensate the influence of ambient temperature variation?

2

Performance Characteristics

To make a decisive choice out of a number of available designs, or to venture on a new instrument design it is very necessary to know the response of the instrument to a particular input. This subject is usually discussed under the following two heads:

(i) Static performance characteristics, and
(ii) Dynamic performance characteristics

2.1 Static performance characteristics

Some applications involve the measurement of quantities which are either constant or vary very slowly with time. The quality of instrument for such a case is determined by its static calibration and by the input-output relationship.

Static Calibration

Static calibration refers to a situation where an input is applied to an instrument and corresponding output measured, while all other inputs (desired, modifying, interfering) are kept constant at some value. The input-output relation developed this way comprises a calibration valid under the stated constant conditions of all other inputs. A family of calibration curves can be obtained by varying any desired input and noting the corresponding outputs, while maintaining all other inputs constant. The static performance characteristics are obtained from these calibration curves.

As a rule calibration standard must be at least ten times more accurate than the instrument to be calibrated.

(i) *Static sensitivity*

The static sensitivity is defined as the slope of the calibration curve, i.e.

$$\text{Sensitivity} = \frac{\Delta q_0}{\Delta q_i}$$

where q_0 and q_i are the output and input, respectively. If the input-output relation is not linear, the sensitivity varies with the input value and is defined as (Fig. 2.1).

$$\text{Sensitivity} = \left.\frac{\Delta q_0}{\Delta q_i}\right|_{q_i}$$

While the instrument's sensitivity to its desired input is of primary importance, its sensitivity to interfering, modifying inputs may also be of some

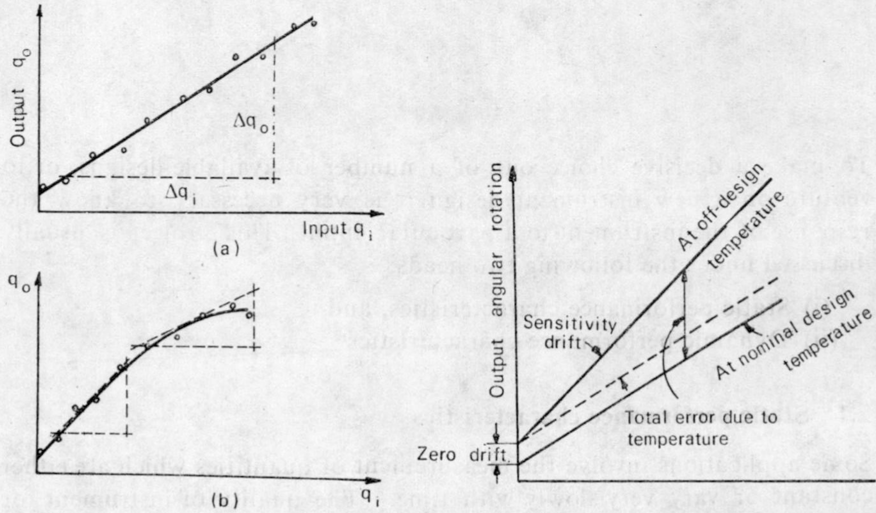

Fig. 2.1 Definition of static sensitivity

Fig. 2.2 Zero and sensitivity drift

interest. Let us consider the case of strain gauge. The temperature is an interfering input and causes the resistance of the gauge to vary and thus would drift the value even when the strain is zero. This is called a zero drift. Further, the temperature is also a modifying input which changes the sensitivity factor. This effect is called as sensitivity drift or scale factor drift. Fig. 2.2 illustrates these points clearly.

(ii) *Linearity*

An instrument may have a linear relation between input and its output. Linearity, however, is never completely achieved and the deviations from the ideal is termed as linearity tolerances. In commercial instruments, the maximum departure from linearity is often specified. The two ways of specifying linearity, namely, independent and proportional linearity are illustrated in Fig. 2.3(a) and (b). In Fig. 2.3(a) 2% linearity means that the output will be within two parallel lines spaced ±2% of the full scale output from the idealised line. In Fig. 2.3(b) 2% proportional linearity is specified. It means that the true point is never more than 2% away from the recorded input regardless of the magnitude of the input. It may be noted that an instrument which does not possess linearity can still be highly accurate.

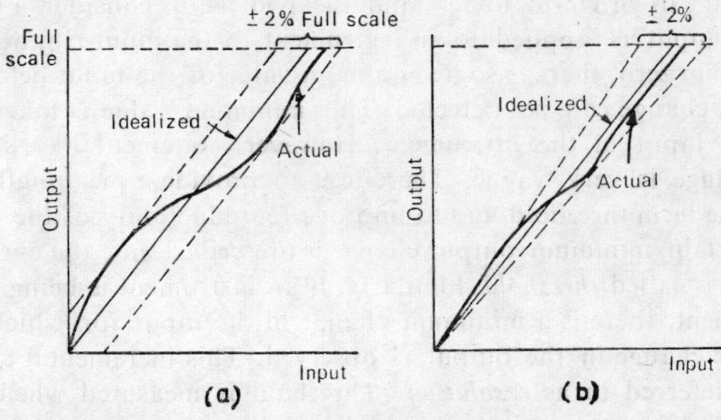

Fig. 2.3 Specifying independent and proportional linearity

(iii) *Repeatability*

If an instrument is used to measure identical input many times, the output is not the same but shows a scatter or deviation. This deviation from the ideal one, in absolute units or a fraction of the full scale, is called repeatability error and is illustrated in Fig. 2.4.

(iv) *Hysteresis—threshold—resolution*

When testing for repeatability, it is often seen that input-output graphs do not coincide when:

(a) the input is continuously increased from zero value, and
(b) the input is continuously decreased from maximum value.

This non-coincidence of input-output graphs arises due to the phenomenon of hysteresis (Fig. 2.5). Some causes of hysteresis effects in an

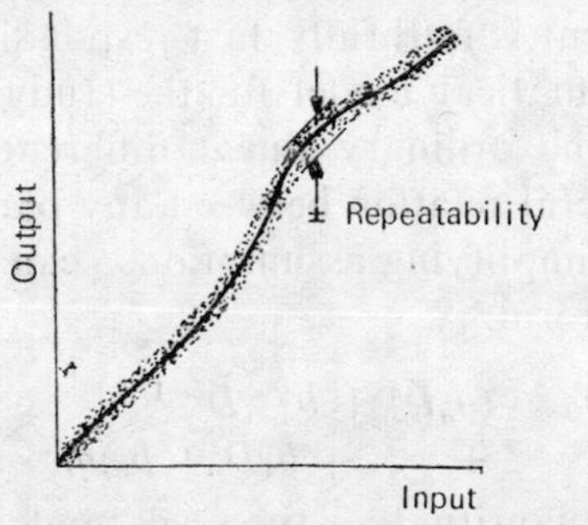

Fig. 2.4 Repeatability error

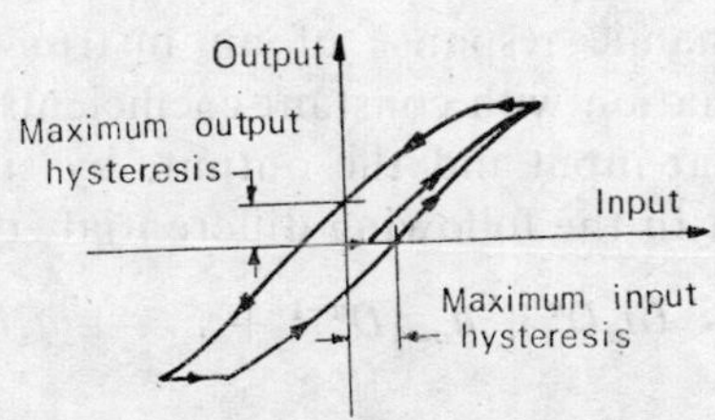

Fig. 2.5 Hysteresis effect

instrument are internal friction, sliding or external friction, free play or looseness of a mechanism. Hysteresis effects are best eliminated by taking readings both for ascending and descending values of the input, and then taking an arithmetic average.

Threshold and resolution are the two characteristics of a measuring instrument. In order to understand the two let us consider a situation when an input is applied to an instrument. If the input is gradually increased from zero, there is some minimum value of the input below which no output change can be detected. This minimum value is taken as the threshold input of the instrument. However, statement like detectable input change is very vague. Therefore, to provide a meaningful definition of the term threshold and to improve reproducibility of the measurement, certain minimum output change is prescribed and the corresponding input is called *threshold*. Similarly, if an instrument is being used for measurement, there is a minimum change in the input for which certain detectable change in the output is observed. This incremental change in input is referred to as *resolution*. Threshold is measured when the input is varied from zero while the resolution is measured when the input is varied from any arbitrary but non-zero value.

It will be observed that the output does not change at all until a certain input increment is exceeded. This increment is called resolution.

Therefore, we can state that resolution defines the smallest measurable input change, while the threshold is a smallest measurable input. Both threshold and resolution can be given either as absolute values or as percentage of the full scale deflection.

(v) *Readability and span*

The readability depends on both instrument and observer for analog display and often not stated. The span refers to the range of the instrument.

2.2 Dynamic performance characteristics

When an instrument is used to measure fast time varying quantities or transient events, it is essential to know how it responds to them. A good dynamic response means that the instrument is faithfully fast responding as the input varies. A most useful mathematical model for the study of dynamic response of an instrument is the ordinary linear differential equation with constant coefficients. A general relation between any particular input and the output, by suitable simplifying assumptions, can be put in the following differential equation form:

$$(a_n D^n + a_{n-1} D^{n-1} + \dots + a_1 D + a_0) q_0 = (b_n D^n + b_{n-1} D^{n-1} + \dots + b_1 D + b_0) q_i$$

where q_0 = output quantity, q_i = input quantity, t = time, a's and b's are physical constants and $D = d/dt$ differential operator.

The task is to find out q_0 in terms of system parameters and input, q_i.

Solution of the above differential equation can be written as

$$q_0 = q_{0cf} + q_{0pi}$$

where q_{0cf} = complimentary function part of solution,

q_{0pi} = particular integral part of solution.

The complimentary function part is a solution when no input is applied. This gives the natural behaviour of the system. The particular integral part is the solution due to the impressed input.

The ways and means of solving a general differential equation will not be discussed, but the solutions to particular cases as and when they arise will be discussed. However, it will be worthwhile to bring in here the concept of transfer function and frequency response.

(i) *Operational transfer function*

The operational transfer function relating the output q_0 to input q_i is defined by the equation

$$\frac{q_0}{q_i}(D) = \frac{b_n D^n + b_{n-1} D^{n-1} + \ldots + b_1 D + b_0}{a_n D^n + a_{n-1} D^{n-1} + \ldots + a_1 D + a_0}$$

The representation $\frac{q_0}{q_i}(D)$ is a very general relation. The operational transfer function is a very useful concept in the analysis, design and application of instrument. The merits of operational transfer function approach are:

(a) The dynamic characteristics of the system can be represented by means of block diagram.

(b) It is very helpful to determine the overall characteristics of the system in terms of the transfer functions of its components. If the loading effect is negligible, then the overall transfer function is the product of the transfer functions of the individual components.

(ii) *Sinusoidal transfer function and frequency response*

If an input of the form

$$q_i = A_i \sin \omega t$$

is applied to an instrument and when all the transient effects die out, it will be seen that the output q_0 will be a sine wave of the same frequency (ω) as that of the input. However, the amplitude of output may differ from that of the input, and a phase shift may be present. Since the frequency is the same, the relation between the input and output is completely described by giving their amplitude ratios and phase shift. Both these quantities vary, in general, with the frequency ω. Thus the frequency response of a linear system consists of curves of amplitude ratios and phase shift as a function of frequency.

The sinusoidal transfer function of an instrument is defined as

$$\frac{q_0}{q_i}(i\omega) = \frac{b_n(i\omega)^n + b_{n-1}(i\omega)^{n-1} + \ldots + b_1(i\omega) + b_0}{a_n(i\omega)^n + a_{n-1}(i\omega)^{n-1} + \ldots + a_1(i\omega) + a_0}$$

The quantity $\frac{q_0}{q_i}(i\omega)$ is a complex quantity. Its magnitude $\left|\frac{q_0}{q_i}(i\omega)\right|$ gives the amplitude ratio while its argument is the phase. The relationship between this complex quantity and frequency constitutes the frequency response.

2.3 Input types

Although a measured quantity, in general, will not be a simple function of time, much can be learnt about an instrument by observing its response to some elementary inputs. The four types of elementary inputs are considered here:

(i) STEP INPUT

It is mathematically represented as

$$q_i = 0 \quad \text{at } t < 0$$
$$= q_{is} \text{ at } t \geqslant 0$$

where q_{is} is a constant value. This is illustrated in Fig. 2.6(a).

(ii) RAMP INPUT

It is mathematically represented as

$$q_i = 0 \qquad \text{at } t < 0$$
$$q_i = t \cdot q_{is} \quad \text{at } t \geqslant 0$$

where q_{is} is the slope of input vs. time relation. This is illustrated graphically in Fig. 2.6(b).

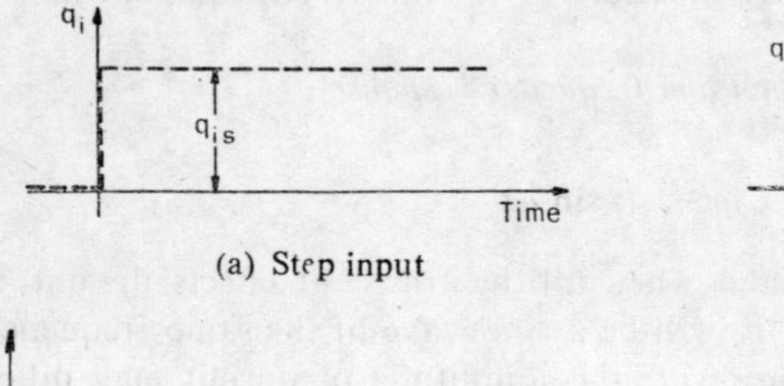

(a) Step input

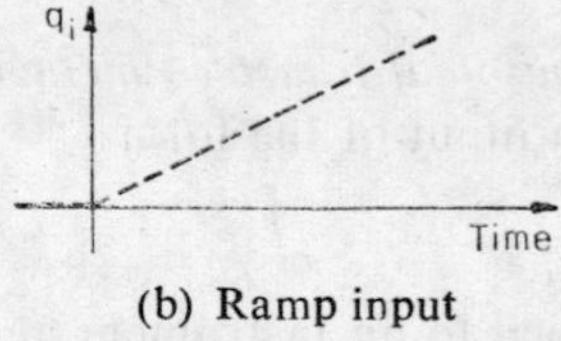

(b) Ramp input

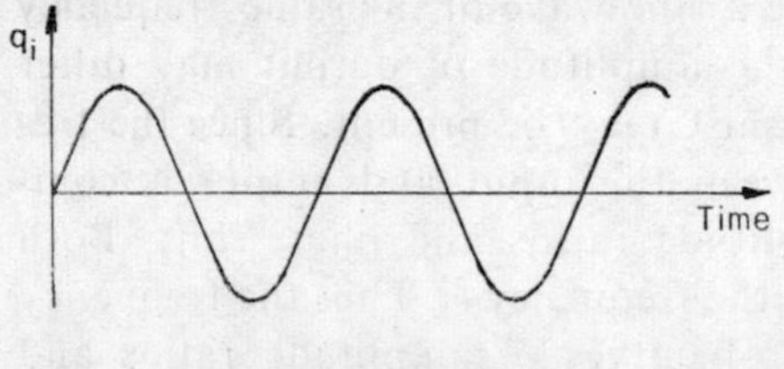

(c) Sinusoidal input

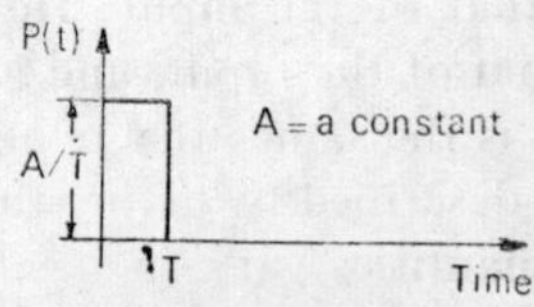

(d) Impulse input

Fig. 2.6

(iii) SINE INPUT

This is represented by

$$q_i = A_i \sin \omega t$$

where A_i is the amplitude and ω its frequency. This is one of the very important elementary inputs and is shown in Fig. 2.6(c).

(iv) IMPULSE INPUT

The impulse function of strength A is defined by $\underset{T\to 0}{\triangleq} p(t)$ and is illustrated in Fig. 2.6 (d). It has the following properties:

(a) The duration of pulse is infinitesimal.
(b) The peak of the pulse is infinitely high.
(c) The area of the pulse is finite and is equal to the strength of the pulse, A.

If the area of the pulse is unity, it is called as unit impulse function $u_1(t)$. Thus $p(t) = Au_1(t)$ (Fig. 2.6(d)). The impulse function is very useful in studying the frequency response of mechanical instruments as it possesses all the frequency components. This function is a Dirac-delta function.

2.4 Instrument types

Although the general treatment outlined earlier is adequate for handling any linear measurement system, some of the instruments warrant separate treatment. Further, some of the salient features would be pointed out while considering the individual cases.

(i) ZERO ORDER INSTRUMENT

The differential equation which describes the zero order instrument is given by

$$q_0 = \frac{b_0}{a_0} q_i = kq_i$$

where $k = \dfrac{b_0}{a_0}$ is the static sensitivity of the instrument. The block diagram representation of zero order instrument is given in Fig. 2.7(a).

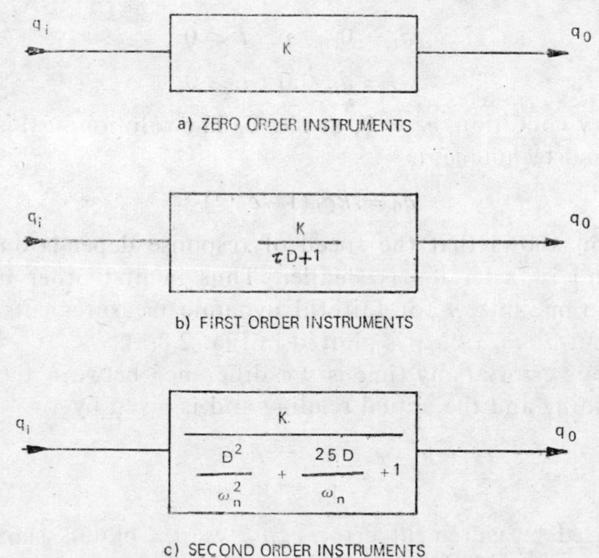

Fig. 2.7 Block diagram representation

It is obvious from the above equation that no matter how q_i might vary with time, the output follows it perfectly with no distortion and time lag. The dynamic response of zero order instrument is ideal or perfect. Some examples of this include mechanical lever, linear electrical potentiometer, amplifier, etc.

(ii) First Order Instrument

If in the general differential equation, all the constants except a_i, a_0 and b_0 are zero, one gets

$$\left(\frac{a_1}{a_0} D + 1\right) q_0 = \frac{b_0}{a_0} q_i$$

or

$$(\tau D + 1) q_0 = k q_i$$

where $\dfrac{a_1}{a_0} = \tau$ is called time constant and k is the static sensitivity. An instrument which follows this equation is a first order instrument. The operational transfer function of first order instrument is, therefore, given by

$$\frac{q_0}{q_i}(D) = \frac{k}{\tau D + 1}$$

The block diagram representation is given in Fig. 2.7(b).

(a) *Step response of a first-order instrument*

In order to study the step response solve the differential equation

$$\frac{q_0}{q_i}(D) = \frac{k}{\tau D + 1}$$

under the initial conditions

$$q_i = 0 \quad \text{at} \quad t < 0$$

$$= q_{is} \quad \text{at} \quad t \geqslant 0$$

and boundary condition, $q_0 = 0$, at $t < 0$. The solution of this equation, using the usual technique, is

$$q_0 = k q_{is}(1 - e^{-t/\tau})$$

This equation shows that the speed of response depends only on the value of τ and is faster if τ is smaller. Thus in first order instruments one desires to minimize τ for faithful dynamic measurements. The normalised output q_0/kq_i vs. t/τ is plotted in Fig. 2.8(a).

The dynamic error at any time is the difference between the ideal (no time lag) reading and the actual reading and is given by

$$e_m = q_{is} - q_0/k$$

$$= q_{is} e^{-t/\tau}$$

The normalised measurement error e_m/q_{is} vs. t/τ plot is shown in Fig. 2.8(b). The output, of a first order instrument after the application of a step input, rises and after a very long time reaches the final value. The

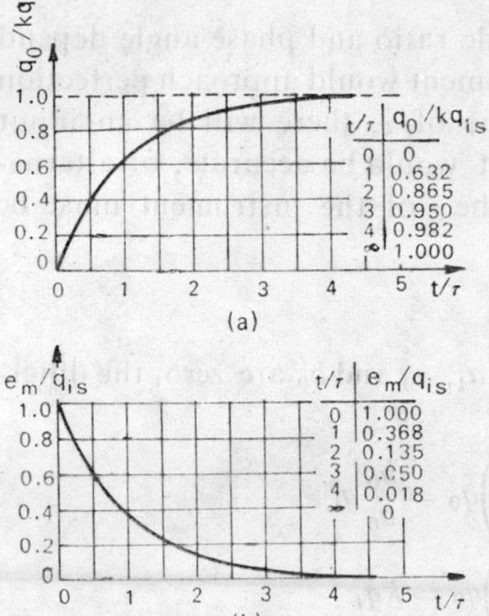

Fig. 2.8 Non-dimensional step-function response

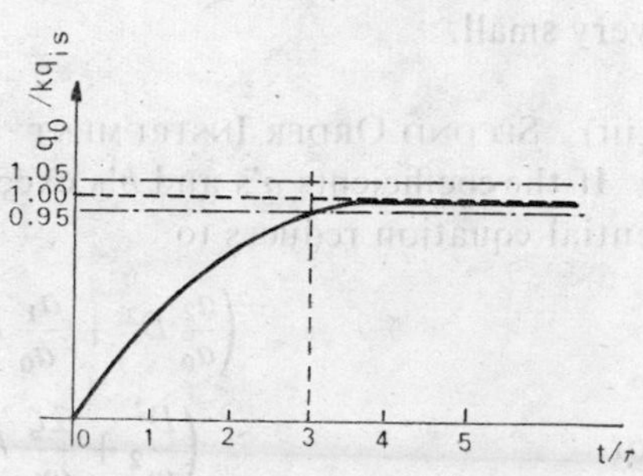

Fig. 2.9 Definition of setting time

dynamic characteristic of the instrument is given by settling time. A small settling time is indicative of fast response of the instrument. The settling time is defined as the time, after the application of step input, for the output to reach and stay within a specified tolerance band about the final value. Therefore, its numerical value will depend on the percentage tolerance band used. Often one speaks of 5% settling time. A 5% settling time for a first order instrument is equal to three times its time constant as shown in Fig. 2.9. Other percentages can also be used.

(b) *Frequency-response of a first order instrument*

This requires the solution of differential equation when an input of the form

$$q_i = A_i \sin \omega t$$

is applied. The initial conditions are:

$$q_i = q_0 = 0, \quad \text{at } t < 0$$
$$q_i = A_i \sin \omega t \text{ at } t \geqslant 0$$

The solution, under these initial conditions, is given by

$$q_0 = ce^{-t/\tau} + \frac{kq_i}{1 + i\omega\tau}; \text{ here } c \text{ is a constant.}$$

As the time flows, the complimentary function solution dies out and hence the steady state solution is

$$\frac{q_0}{kq_i} = \frac{1}{1 + i\omega\tau} = \frac{1}{\sqrt{1 + \omega^2\tau^2}} \angle -\tau\omega$$

Thus the amplitude ratio is given by $1/\sqrt{1+\omega^2\tau^2}$ and phase angle by $\tan^{-1}(-\tau\omega)$.

Thus it is obvious that both amplitude ratio and phase angle depend on both ω and τ. Thus a first order instrument would approach perfection if the value of $\omega\tau$ is small. For any value of τ, there will be an input frequency below which the measurement would be accurate, or alternatively, for high frequency measurement, the τ of the instrument must be very small.

(iii) SECOND ORDER INSTRUMENT

If the coefficients a's and b's except a_2, a_1, a_0 and b_0 are zero, the differential equation reduces to

$$\left(\frac{a_2}{a_0} D^2 + \frac{a_1}{a_0} D + 1\right) q_0 = \frac{b_0}{a_0} q_i$$

or

$$\left(\frac{D^2}{\omega_n^2} + \frac{2\zeta}{\omega_n} D + 1\right) q_0 = k q_i$$

where $\omega_n = \sqrt{a_0/a_2}$ = undamped natural frequency, rad/sec,

$$\zeta = \frac{a_1}{2\sqrt{a_0 a_2}} = \text{damping ratio, dimensionless, and}$$

$$k = b_0/a_0 = \text{static sensitivity.}$$

An instrument which follows this equation is termed as second order instrument.

The operational transfer function of a second order instrument is given by

$$\frac{q_0}{q_i}(D) = \frac{k}{D^2/\omega_n^2 + 2\zeta D/\omega_n + 1}.$$

The block diagram representation is given in Fig. 2.7(c). Some examples of second order instruments are force measuring spring scale, spring mass systems under impressed force, piezoelectric pick-ups, recorder, etc.

(a) *Step response of a second order instrument*

To study the step response of a second order system, consider the differential equation

$$\left(\frac{D^2}{\omega_n^2} + \frac{2\zeta D}{\omega_n} + 1\right) q_0 = k q_i$$

with a set of initial conditions

$$q_i = q_0 = 0 \quad \text{at } t < 0$$

and $$q_i = q_{is} \quad \text{at } t \geqslant 0$$

Further $$dq_0/dt^0 = 0 \quad \text{at } t = 0$$

Solving the above differential equation the following three solutions are obtained corresponding to the three cases:

$\zeta < 1$: UNDERDAMPED SYSTEM

$$\frac{q_0}{kq_{is}} = 1 - \frac{\exp(-\zeta\omega_n t)}{\sqrt{1-\zeta^2}} \sin(\sqrt{1-\zeta^2}\omega_n t + \phi)$$

where $\sin\phi = \sqrt{1-\zeta^2}$.

$\zeta = 1$: CRITICALLY DAMPED SYSTEM

$$\frac{q_0}{kq_{is}} = 1 - (1 + \omega_n t)\exp(-\omega_n t)$$

$\zeta > 1$: OVERDAMPED SYSTEM

$$\frac{q_0}{kq_{is}} = 1 - \frac{\zeta + \sqrt{\zeta^2 - 1}}{2\sqrt{\zeta^2 - 1}} \exp\{(-\zeta + \sqrt{\zeta^2 - 1})\omega_n t\} + \frac{\zeta - \sqrt{\zeta^2 - 1}}{2\sqrt{\zeta^2 - 1}} \exp\{(-\zeta - \sqrt{\zeta^2 - 1})\omega_n t\}$$

The response of a second order system to a step input is plotted in Fig. 2.10 for different values of ζ. It is clear that an increase in the value of ζ reduces the oscillations but slows the speed of response as well. A designer can select a proper value of ζ depending on the requirement of settling time. However, the situation is complicated by the fact that actual form

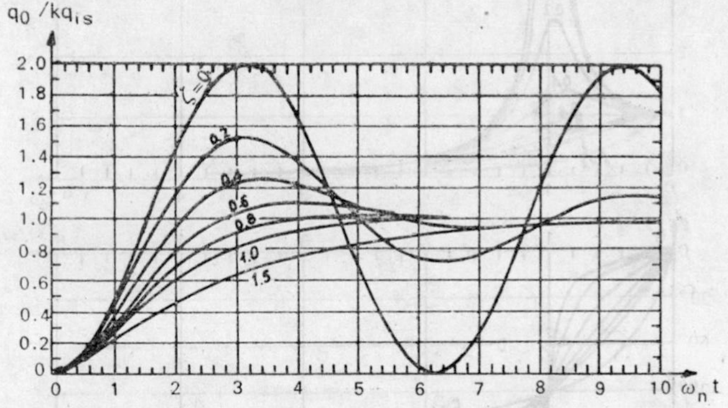

Fig. 2.10 Normalised step response of a 2nd order system

of the input is very complicated and its actual form influences the best value of ζ. Therefore, for inputs of variable and complicated forms, some compromise should be struck. It will be found that most commercial instruments use $\zeta = 0.6$ to 0.7. It will be shown that this range of ζ values gives good frequency response over the widest frequency range.

(b) *Sinusoidal response of second order system*

When a sinusoidal input of the form $q_i = A \sin \omega t$ is applied to the system, the steady state solution of the differential equation represents a sinusoidal output with different amplitude and phase shift but of same frequency. The sinusoidal transfer function of a second order instrument is, therefore, written as:

$$\frac{q_0}{q_i}(i\omega) = \frac{k}{\left(\frac{i\omega}{\omega_n}\right)^2 + \frac{2\zeta i\omega}{\omega_n} + 1}$$

which can be put in the form

$$\frac{q_0/k}{q_i}(i\omega) = \frac{1}{\left[\left(1-\left(\frac{\omega}{\omega_n}\right)^2\right)^2 + \frac{4^2\omega^2}{\omega_n^2}\right]^{1/2}} \angle\phi; \quad \text{where } \tan\phi = \frac{2\zeta}{\left(\frac{\omega}{\omega_n} - \frac{\omega_n}{\omega}\right)}$$

Figure 2.11 illustrates the plots of q_0/kq_i vs. ω/ω_n for various values of ζ. The occurrence of resonance is seen here in the vicinity of $\omega = \omega_n$, i.e., where the forcing frequency ω is equal to the natural frequency ω_n. Extreme amplitudes will be reached unless either damping is introduced or the

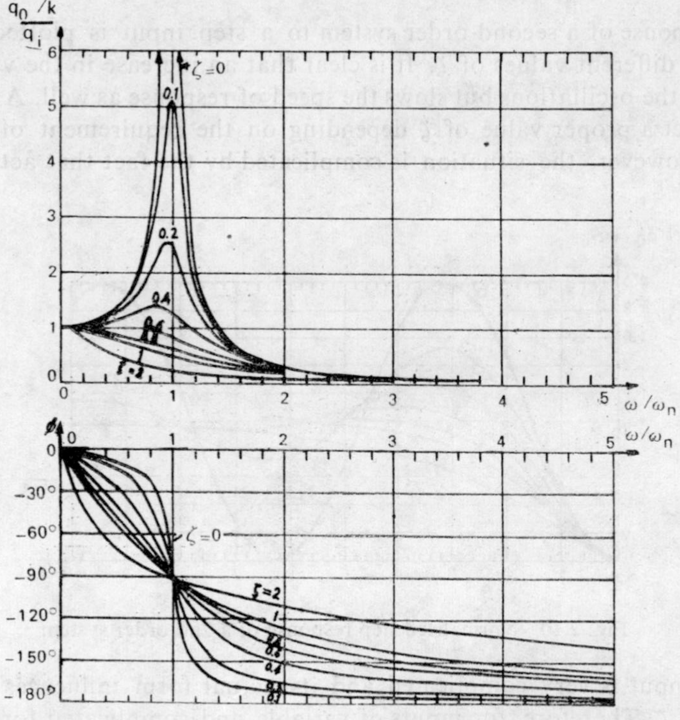

Fig. 2.11 Frequency response of order system

system operates at $\omega = \omega_n$ for a very short time. Further as ω_n increases, the range of frequencies for which amplitude ratio curve is relatively flat also increases; thus a high value of ω_n is needed to measure high frequency q_i's accurately. The widest flat amplitude ratio exists for ζ of about 0.6 to 0.7. While zero phase angle is ideal, it is rarely possible to realise it even approximately.

2.5 Experimental determination of system parameters

The determination of the order of the instrument and then the methods to measure the parameters characterising the system are now discussed with respect to elementary inputs: step input and sinusoidal input.

(a) Zero Order Instrument

It is seen that zero order instrument is a perfect instrument and its dynamical response is ideal. It has no lag or distortion. The constant k, static sensitivity, is the only constant which characterises it and can be obtained by the process of static calibration.

(b) First Order Instrument

The first order instrument is characterised by two parameters τ and k. The static sensitivity k can again be obtained by static calibration. For the determination of the time constant τ, the following two methods, which employ step input, are described.

One common method is to apply a step input and measure τ as the time required to reach 63.2% of the final value. Analytically

$$\frac{q_0/k}{q_{is}} = 1 - e^{-t/\tau} = 0.632$$

or

$$e^{-t/\tau} = 0.368 = \frac{1}{e}$$

or

$$t = \tau$$

The above method depends only on the measurement at two points, i.e., at $t = 0$ and $t = \tau$ and is influenced by the inaccuracies in the determination of $t = 0$ point. Further, it does not determine the order of the instrument. In the method described below, a step input is applied. Therefore, one has

$$1 - \frac{q_0/k}{q_{is}} = e^{-t/\tau}$$

Defining $z = \log_e (1 - q_0/k/q_{is})$ as the incomplete response, and then plotting z vs. t, one obtains a straight line if the system is first order. The slope of this line is $-1/\tau$. This gives a more accurate value of τ because a best straight line is drawn through all the points. Both these methods are graphically described in Figs. 2.12(a) and (b).

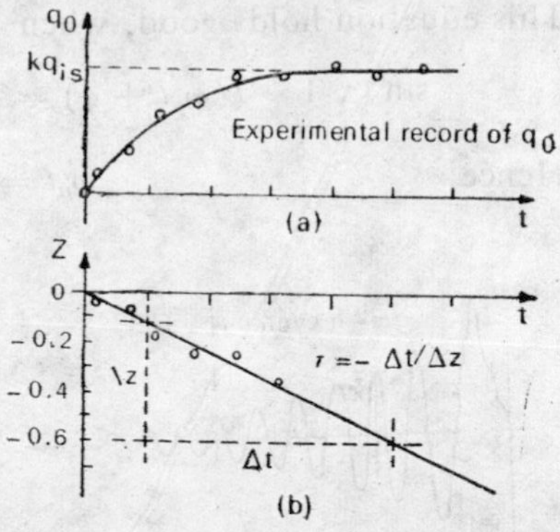

Fig. 2.12 Measurement of τ from step response of order system

(c) Second Order Instrument

The second order instrument is characterised by three parameters, k, ζ and ω_n. The value of k is again obtained by static calibration, while the methods to determine values of ζ and ω_n depend on the value of ζ itself.

The step response of a second order underdamped ($\zeta < 1$) instrument is given by

$$q_0 = A\left[1 - \frac{\exp(-\zeta\omega_n t)}{\sqrt{1-\zeta^2}} \sin(\sqrt{1-\zeta^2}\,\omega_n t + \phi)\right]$$

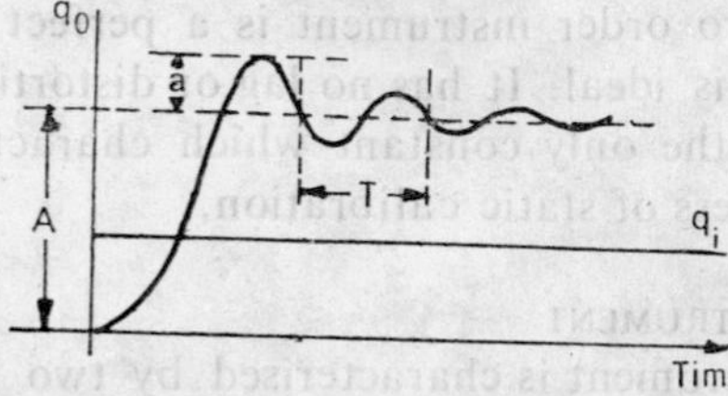

Fig. 2.13 Step response of a 2nd order system for parameter determination

This is plotted in Fig. 2.13. If the time period of damped oscillation is T, we have

$$\sqrt{1-\zeta^2}\omega_n t + \phi + 2\pi = \sqrt{1-\zeta^2}\,\omega_n(t+T) + \phi$$

or

$$\omega_n = \frac{2\pi}{\sqrt{1-\zeta^2}\,T}$$

Thus to obtain ω_n the value of ζ is required. If the overshoot is a, then

$$A + a = A\left[1 + \frac{\exp(-\zeta\omega_n t)}{\sqrt{1-\zeta^2}}\right]$$

or $\quad \dfrac{a}{A} = \exp(-\zeta\omega_n t)$; the denominator $\sqrt{1-\zeta^2} \simeq 1$

This equation holds good, when

$$\sin(\sqrt{1-\zeta^2}\omega_n t + \phi) = -1, \text{ with } \sin\phi = \sqrt{1-\zeta^2} \simeq 1$$

Hence

$$\omega_n t = -\frac{\pi}{\sqrt{1-\zeta^2}}$$

Therefore, $\quad \dfrac{a}{A} = \exp\left(-\dfrac{\zeta\pi}{\sqrt{1-\zeta^2}}\right)$

giving

$$\zeta = \frac{1}{\left[1 + \left(\pi/\ln\dfrac{a}{A}\right)^2\right]^{1/2}}$$

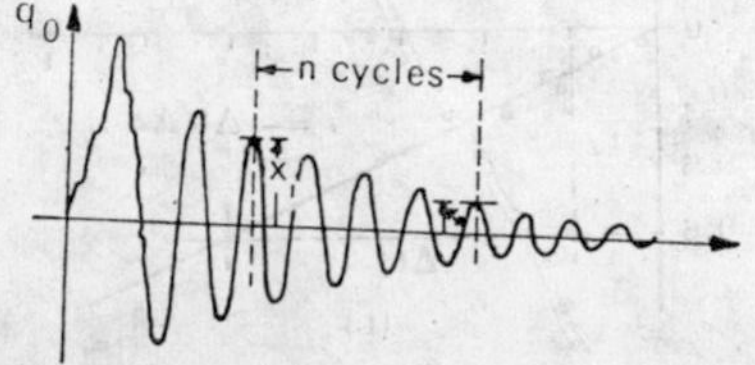

Fig. 2.14 Response of a line order system to a transient input

However, when the system is lightly damped, any fast transient input will produce a response similar to the one shown in Fig. 2.14. The value of τ can be approximated by

$$\zeta = \frac{\log(x_1/x_n)}{2\pi n}$$

where x_1, x_n, n are defined in Fig. 2.14.

Overdamped system ($\zeta > 1$) does not show any oscillations for the step input and hence the determination of ζ and ω_n becomes very difficult. Usually it is easier to express the system performance by two time constants, τ_1 and τ_2 instead of ζ and ω_n. The response of the system to step input is then given by

$$\frac{q_0 k}{q_{is}} = \frac{\tau_1}{\tau_2 - \tau_1} e^{-t/\tau_1} - \frac{\tau_2}{\tau_2 - \tau_1} e^{-t/\tau_2} + 1$$

where $$\tau_1 = \frac{1}{(\zeta - \sqrt{\zeta^2 - 1})\omega_n}$$

and $$\tau_2 = \frac{1}{(\zeta + \sqrt{\zeta^2 - 1})\omega_n}$$

Thus $$\tau_1 > \tau_2$$

To find τ_1 and τ_2 and hence ζ and ω_n from the step response curve one may proceed as follows:

(i) Define a percentage incomplete response R_{p1} as

$$R_{p1} = \left[1 - \frac{q_0/k}{q_{is}}\right] \times 100$$

(ii) Plot R_{p1} on a logarithmic scale vs. time on a linear scale. This curve will approach a straight line for large t if the system is second order. Extend the line back to $t = 0$ and note the value P_1 where this line intersects the R_{p1} scale. Now, τ_1 is the time at which the straight line asymptote has the value $0.368P_1$.

(iii) Now plot on the same graph, a new curve which is the difference between the straight line asymptote and R_{p1}. If this new curve is not a straight line, the system is not second order. If this curve is a straight line the time at which this line has the value $0.368(P_1 - 100)$ is numerically equal to τ_2.

These statements are graphically illustrated in Fig. 2.15. The validity of the above statements and soundness of the procedure can be demonstrated by the following mathematical procedure:

(i) $$R_{p1} = \left[\frac{\tau_2}{\tau_2 - \tau_1} e^{-t/\tau_2} - \frac{\tau_1}{\tau_2 - \tau_1} e^{-t/\tau_1}\right] \times 100$$

(ii) $$\ln R_{p1} = \ln 100 - \frac{t}{\tau_1} + \ln\left[\frac{\tau_2}{\tau_2 - \tau_1} \exp\left\{-t\left(\frac{1}{\tau_2} - \frac{1}{\tau_1}\right)\right\} - \frac{\tau_1}{\tau_2 - \tau_1}\right]$$

since $\tau_1 > \tau_2$ the term $\exp\left\{-t\left(\frac{1}{\tau_2} - \frac{1}{\tau_1}\right)\right\}$ will decrease fast with time t, so for large t,

$$\ln R_{p1s} = \ln 100 - \frac{t}{\tau_1} + \ln\frac{\tau_1}{\tau_1 - \tau_2}$$

The relation between $\ln R_{p1s}$ and t is linear,

or $$R_{p1s} = 100 \frac{\tau_1}{\tau_1 - \tau_2} e^{-t/\tau_1}$$

The value of R_{p1s} at $t = 0$ is P_1, therefore

$$R_{p1s}\Big|_{t=0} = P_1 = \frac{100\tau_1}{\tau_1 - \tau_2}$$

Hence

$$R_{p1s} = P_1 e^{-t/\tau_1}$$

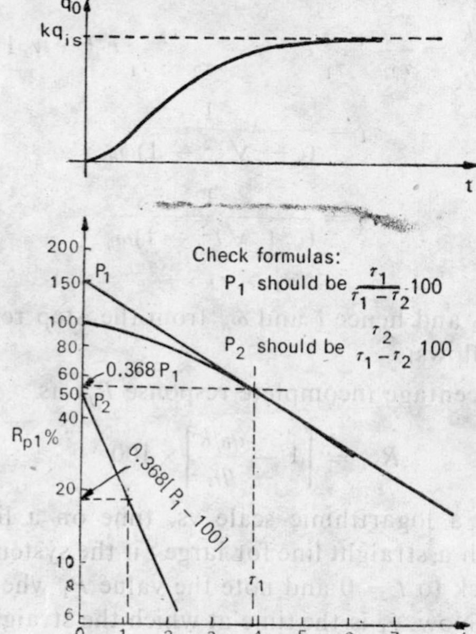

Fig. 2.15 Step test for over damped 2nd order system

Therefore, at $t = \tau_1$ the value of $R_{p1s} = 0.368p_1$.

(iii) Further

$$R_{p1} = 100e^{-t/\tau_1}\left[\frac{\tau_2}{\tau_2 - \tau_1}\exp\left\{-t\left(\frac{1}{\tau_2} - \frac{1}{\tau_1}\right) - \frac{\tau_1}{\tau_2 - \tau_1}\right\}\right]$$

and

$$R_{p1s} = 100\,\frac{\tau_1}{\tau_1 - \tau_2}\,e^{-t/\tau_1}$$

Thus

$$R_{p1s} - R_{p1} = 100\,\frac{\tau_2}{\tau_1 - \tau_2}\,e^{-t/\tau_2}$$

But,

$$R_{p1s} = R_{p1}\Big|_{t=0} = 100\,\frac{\tau_2}{\tau_1 - \tau_2} = P_2 = P_1 - 100$$

So, at $t = \tau_2$, the value of $R_{pis} - R_{pi}$ is $0.368P_2$ or $0.368(P_1 - 100)$.

2.6 Frequency response methods

The frequency response methods are very simple and easy to perform but are very expensive due to the non-availability of mechanical sine generator. Often impulse input response is used to obtain the parameters.

(a) FIRST ORDER INSTRUMENT

The frequency response of a first order instrument is given by

$$\frac{q_0}{q_i}(i\omega) = \frac{k}{\sqrt{\tau^2\omega^2 + 1}} < \phi, \text{ where } \tan\phi = -\omega\tau.$$

It is represented in Fig. 2.16. The asymptote to the curve at high frequency side makes an angle of -20 db/decade. Thus if the slope is -20 db/decade, the instrument is first order. The asymptotes of low and high frequency sides of the curve intersect at a point which corresponds to a frequency such that $\omega_b = \frac{1}{\tau}$.

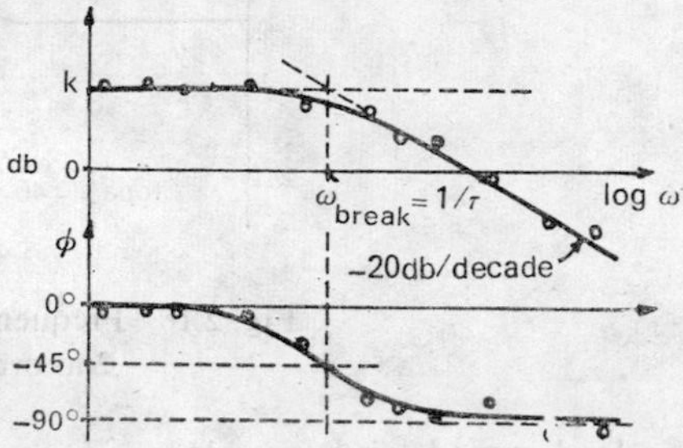

Fig. 2.16 Frequency response test of 1st order system

Thus ω_b can be obtained from measurement and τ can be determined. It is to be noted that ω_b corresponds to a phase angle of $-45°$.

(b) SECOND ORDER INSTRUMENT

The frequency response of a second order system is given by

$$\frac{q_0}{q_i}(i\omega) = \frac{k}{\left\{\left[1-\left(\frac{\omega}{\omega_n}\right)^2\right]^2 + 4\zeta^2\left(\frac{\omega}{\omega_n}\right)^2\right\}^{1/2}} \angle\phi$$

where
$$\tan\phi = \frac{2\tau}{\left(\frac{\omega}{\omega_n} - \frac{\omega_n}{\omega}\right)}$$

$\zeta < 1$: UNDERDAMPED. The output, under this case, is shown in Fig. 2.17. The value of ζ is obtained by solving the equation

$$A_r/A_0 = \frac{1}{2\zeta\sqrt{1-\zeta^2}}$$

where A_r and A_0 are obtained experimentally. They are defined in Fig. 2.17.

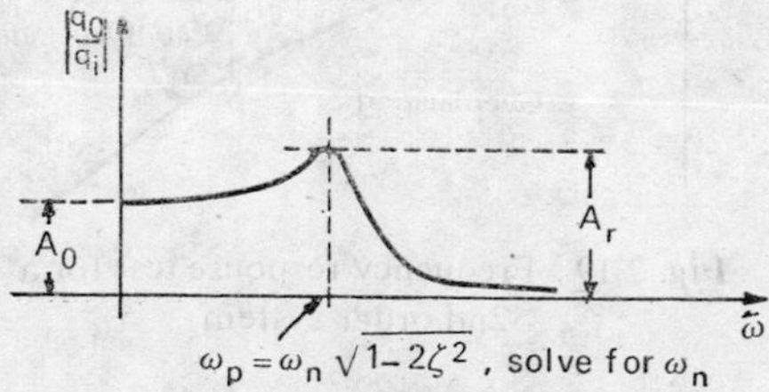

Fig. 2.17 Frequency response test for a 2nd order system

The value of ω_n is obtained by observing the maximum output for the frequency ω_p, where

$$\omega_p = \omega_n\sqrt{1-2\zeta^2}$$

Substitution of the value of ζ in the above equation gives the value of ω_n.

Instead of plotting output vs. frequency, one can plot db vs. log ω. Fig. 2.18 shows a plot of db vs. log ω. This method is identical both for

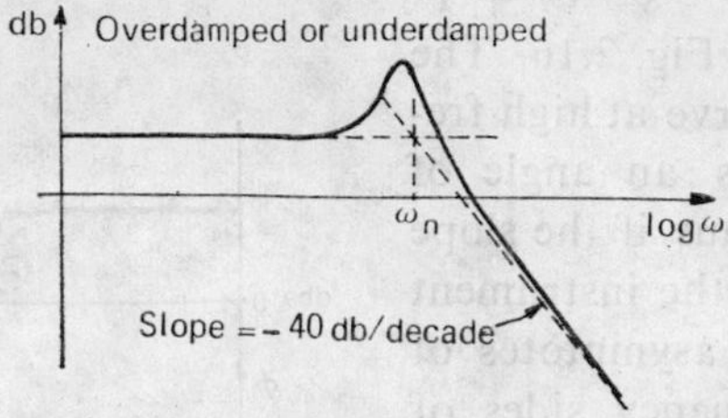

Fig. 2.18 Frequency response test for a 2nd order system

underdamped and overdamped cases. The slope of the straight line drawn asymptotically to the curve at the high frequency side is -40 db/decade. Further the intersection of asymptotic lines at low and high frequency sides of the curve gives ω_n as shown in the figure.

$\zeta = 1$: CRITICALLY DAMPED. In this case, as well, the plot between db and log ω can be used to determine ω_n. The value of ω_n corresponds to the intersection of two asymptotic lines at the low and high frequency side of the curve.

$\zeta > 1$: OVERDAMPED. The plot of db vs. log ω for an overdamped system is shown in Fig. 2.19. There are two breaks in frequencies corresponding to

$$\omega_1 = \frac{1}{\tau_1}$$

and

$$\omega_2 = \frac{1}{\tau_2}$$

This is shown explicitly in Fig. 2.19.

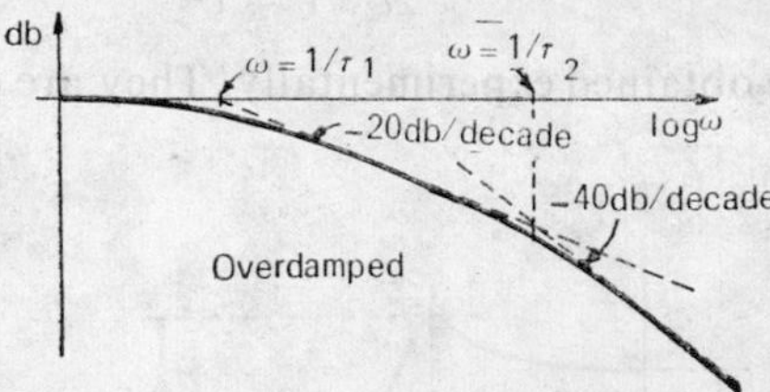

Fig. 2.19 Frequency response test for a 2nd order system

Exercises

1. Prove that the response of a first order instrument to a ramp input is given by

$$\frac{q_0/k}{q_{is}\tau} = \frac{t}{\tau} - (1 - e^{-t/\tau})$$

Show that the instrument always reads what the input was τ seconds before.

2. If a parabolic input of the form $q_1 = q_{is}t^2$ is applied to a second order instrument, prove that its response is given by

$$\frac{q_0/k}{q_{is}\tau^2} = \left(\frac{t}{\tau}\right)^2 - 2(e^{-t/\tau} + \frac{t}{\tau} - 1)$$

3. Prove that the impulse response of a first order instrument is dependent only on its strength and not on the shape and is always given by

$$q_0 = \frac{KA}{\tau} e^{-t/\tau}$$

where A is the strength or impulse defined as

$$A = \int_0^{\Delta t} u(t)\, dt \quad t \to 0;\ u(t) \to \infty$$

4. A part of a continuous sinusoidal input of frequency ω is applied to a thermocouple of time constant τ. Assuming $\omega\tau$ to be very small, obtain the response of the thermocouple. If a single temperature pulse of sinusoidal shape (half wave) of duration t ($t = 4\tau$) is applied to this thermocouple, what will be its response? Compare these two responses.
5. A thermocouple of time constant τ is subjected to an input temperature pulse of maximum value T_0 and duration 7τ as shown in Fig. 2.20. Obtain the output of the

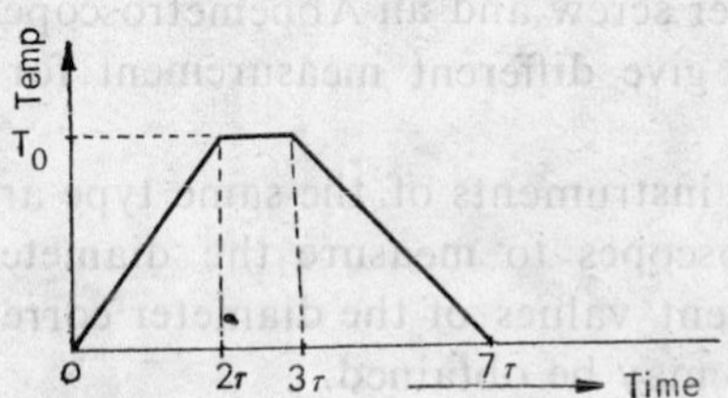

Fig. 2.20 Temperature input to a thermocouple

thermocouple as a function of time. If the duration of the pulse is one-tenth of the time constant, what difference in output is observed? Assume first order response of the thermocouple.

6. Show that the ramp response of second order instrument is given as follows:

$\zeta < 1$:

$$\frac{q_0/k}{q_{is}} = t - \frac{2\zeta}{\omega_n}\left[1 - \frac{\exp(-\zeta\omega nt)}{2\zeta\sqrt{1-\zeta^2}} \sin(\sqrt{1-\zeta^2}\ \omega_n t + \phi)\right]$$

where $\tan\phi = \dfrac{2\zeta\sqrt{1-\zeta^2}}{2\zeta^2 - 1}$.

$\zeta = 1$:

$$\frac{q_0/k}{q_{is}} = t - \frac{2}{\omega_n}\left[1 - \left(1 + \frac{\omega_n t}{2}\right)\exp\left(-\omega_n t\right)\right]$$

$\zeta > 1$:

$$\frac{q_0/k}{q_{is}} = t - \frac{2\zeta}{\omega_n}\left[1 + \frac{2\zeta^2 - 1 - 2\zeta\sqrt{\zeta^2-1}}{4\zeta\sqrt{\zeta^2-1}} \exp\{(-\zeta + \sqrt{\zeta^2-1})\omega_n t\}\right.$$
$$\left. + \frac{-2\zeta^2 + 1 - 2\zeta\sqrt{\zeta^2-1}}{4\zeta\sqrt{\zeta^2-1}} \exp\{(-\zeta - \sqrt{\zeta^2-1})\omega_n t\}\right]$$

3

Inaccuracy of Measurement and Its Analysis

3.1 Introduction

The object of each measurement is to describe some physical property of an object quantitatively: length, temperature, pressure, etc. Every measurement of such a quantity has a certain amount of uncertainty. This may be explained by considering the following experiments in measuring the diameter of a disc:

(i) First, three different instruments are chosen: a caliper square, a micrometer screw and an Abbemetroscope. Each of these instruments may give different measurement for the diameter of the same disc.

(ii) Next, three instruments of the same type are used, for example, three metroscopes to measure the diameter of the disc. Again three different values of the diameter corresponding to the three instruments may be obtained.

(iii) Lastly, only a single instrument is used to measure the diameter of the disc three times. Here as well one may get three different values.

Thus one finds that even repeated measurements on the same instrument may give different values of the same physical quantity. Therefore, it can be concluded that all measured values are inaccurate to some degree. It is in fact impossible to find the true value, although it is safe to assume that the true value exists. The aim is, therefore, to find the most probable value and assign the uncertainty to it. In other words, the aim of the experimenter is not to make the uncertainty of the measurements as small as possible; a cruder result may serve his purpose well enough, but he must be assured that the uncertainty in his measurement is so small as not to affect his conclusions he draws from his results. The task is to determine how uncertain a measurement may be and devise a consistent way of specifying the uncertainty in an analytical form.

3.2 Types of errors

The errors are, in general, classified as accidental and systematic. Usually each error gives an accidental and a systematic component which should be used in assigning uncertainty to the measured value. Besides, there may be a gross blunder, which arises due to faulty design, faulty circuit, etc., and an experimenter should be able to eliminate this hopefully.

(a) Accidental Error

These accidental errors are random in their incidence and variable in magnitude and usually follow a certain statistical law—usually normal distribution law. A measure of accidental errors is, therefore, standard deviation and with increasing number of measurements of the same quantity, the value of the standard deviation becomes smaller and the value of the physical quantity more sure. These errors are usually introduced either by environmental fluctuations or/and observer's mental state and visual conditions.

(b) Systematic Error

The systematic errors have definite magnitude and direction. These are usually more troublesome, for repeated observations need not necessarily reveal them and even when their nature or existence has been established, they are sometimes very difficult to determine and eliminate. Therefore in most of the cases the systematic errors are not used for correcting the measurements, but taken with their full values as uncertainty along with the accidental errors.

The systematic errors of unknown sign arise due to the production tolera nce. Any component in an instrumental setup will be either more or less than the nominal value. The true value itself and the direction of deviation from the true value are not known. The tolerance values, therefore, may be called systematic errors of unknown sign and they are usually treated same as accidental errors.

The experimenter may sometimes use theoretical methods to estimate the magnitude of systematic errors. For instance, error introduced in the measurement of temperature due to the exposed portion of mercury thermometer can be estimated theoretically.

(c) Mean as a Best Value

When a large number of measurements on a physical quantity are made, all the measurements differ in magnitude from each other and also from the true value which is not known. The task is to obtain the value of the physical quantity which is very close to the true value.

If a quantity x_0 (unknown) is measured n times and recorded as x_1, x_2, . . . , x_n units, then $x_r = x_0 + e_r$, where e_r is the uncertainty in the observation and can take both positive and negative values. The arithmetic mean $\bar{x}$ of the n measurements is

$$\bar{x} = \frac{x_1 + x_2 + \ldots + x_n}{n} = x_0 + \frac{e_1 + e_2 + \ldots + e_n}{n}$$

Since some of the errors are positive and some are negative, the $(e_1 + e_2 + \ldots + e_n)/n$ will be very small. In any case it will be smaller numerically than the greatest value of the separate errors. Thus if e is the largest numerical error in any of the measurements,

$$(e_1 + e_2 + \ldots + e_n)/n \leqslant e$$

and consequently

$$\bar{x} - x_0 \leqslant e$$

Hence in general $\bar{x}$ will be nearer to x_0 and may be taken as the best value of physical quantity. In general, larger the value of n, the more $\bar{x}$ approaches x_0.

It may be noted that it is not possible to find the values of e_r as x_0 is not known. It is, therefore, usual to examine the scatter or dispersion around the mean value $\bar{x}$ rather than x_0. Thus

$$x_r = \bar{x} + d_r = x_0 + e_r$$

or

$$\bar{x} - x_0 = e_r - d_r$$

Thus

$$\Sigma d_r = 0$$

3.3 Accuracy and precision

If an instrument measures 'm' instead of M, then $(M - m)$ or $(M - m)/M$ is a measure of accuracy of the instrument. The experimenter should know the degree of accuracy that he can achieve from his instrument. Accuracy is usually quoted as a percentage figure based on the full scale reading of the instrument. Thus if a pressure gauge* has a range from 0 to 1 kg/cm^2 and quoted inaccuracy of ± 1.0 per cent of the full scale, this is to be interpreted that no error greater than $\pm$ 0.01 kg/cm^2 can be expected for any reading that might be taken on this gauge, provided it is properly used. Note that for an actual reading of 0.1 kg/cm^2 an error of ± 0.01 kg/cm^2 is 10 per cent of the reading.

Another method sometimes used gives the error as a percentage of a particular reading with a qualifying statement to apply to the lower end of the scale. For example, a spring scale might be described as having an i naccuracy of ± 0.5 per cent of the reading or ± 0.05 kgf whichever is greater.

It is worthwhile to note the difference between the terms 'accuracy' and 'precision' and one should be very careful in their usage. Accuracy refers to the *closeness of the measurement to the true value of the physical quantity*, whereas the term precision is used to indicate the *closeness with which the measurements of the same input quantity agree with one another* quite independently of any systematic error involved. Mathematically, if d_r is small, the precision is high and e_r is small, the accuracy of measurement is high. As an example, consider the measurement of a known voltage of 100 volts with a certain meter. Five readings are taken, and the indicated values are 104, 103, 105, 103, 105 volts. It is thus clear that the instrument cannot be depended upon for an accuracy better than 5%,

*Pressure gauges in metric units are normally labelled in kg/cm² meaning kgf/cm².

while a precision of $\pm 1\%$ is indicated because the maximum deviation from the mean value is one volt only. It should be noted that the accuracy of the instrument can be improved by calibration but not better than the precision. Accuracy, therefore, includes precision but the converse is not necessarily true.

3.4 Statistical analysis of data

It has been said earlier that the accidental errors often follow a statistical distribution law. Further, this statistical law, when large number of observations are made, is of Gaussian nature.

In order to build up the necessary theory, assume that a physical quantity is measured n times, and the measurements are $x_1, x_2, \ldots, x_n$, which occur with frequencies $f_1, f_2, \ldots, f_n$, such that $f_1 + f_2 + \ldots + f_n = n$.

The mean value of the measurement is defined as

$$\overline{x} = \frac{\sum_i f_i x_i}{\sum_i f_i}$$

and the variance σ^2 is

$$\sigma^2 = \frac{\sum_i f_i (x_i - \overline{x})^2}{\sum_i f_i}$$

The square root of variance is called the standard deviation σ of the distribution and is a measure of scatter of measurements. However, when the number of observations is not very large, a standard deviation 's' is defined as

$$s = \sqrt{\frac{\sum_i f_i (x_i - \overline{x})^2}{n - 1}}$$

Here n is replaced by $n - 1$ in the denominator and this is known as Bessel's correction.

3.5 Probability concept and distribution law

Let us consider that 162 measurements have been made of a strain in an element using a well proved procedure and apparatus. These measurements are arranged with a range of 0.0025 and are given in Table 3.1 with the frequency of occurrence.

Suppose a quantity z is defined by

$$z = \frac{\text{Number of measurements in an interval} \Big/ \text{Total number of measurements}}{\text{Width of interval}}$$

and a graph with height z for each interval is plotted. Such a plot is called histogram (Fig. 3.1). The area under every rectangle is numerically

TABLE 3.1 Results of a strain measurement

Range of strain values	No. of readings within the range
0.0800–0.0825	1
0.0825–0.0850	1
0.0850–0.0875	0
0.0875–0.0900	2
0.0900–0.0925	2
0.0925–0.0950	4
0.0950–0.0975	9
0.0975–0.1000	22
0.1000–0.1025	18
0.1025–0.1050	25
0.1050–0.1075	23
0.1075–0.1100	15
0.1100–0.1125	7
0.1125–0.1150	12
0.1150–0.1175	8
0.1175–0.1200	8
0.1200–0.1225	2
0.1225–0.1250	2
0.1250–0.1275	1
Total	162

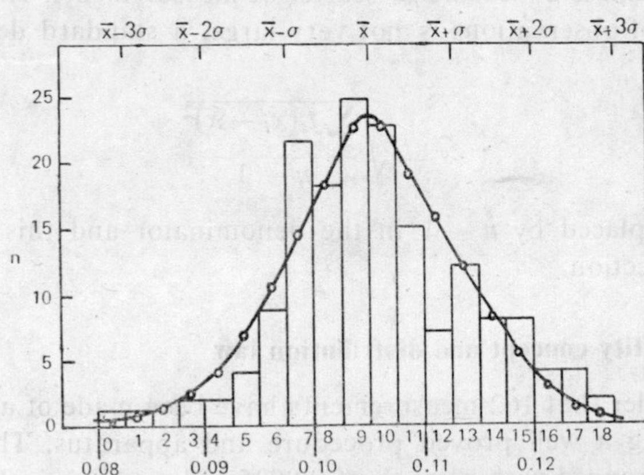

Fig. 3.1 Histogram of strain measurement and superposed normal distribution

equal to the probability that a particular measurement will fall in that interval. When an infinite number of measurements are made and the intervals are made as small as possible, the plot would approach a limiting case of a smooth curve. If this limiting case is taken as a mathematical model of real physical situation, the function $z = f(x)$ is called the probability density function for the mathematical model of a real physical process. Thus the probability of a measurement lying between a and b (Fig. 3.2(a)) will be given by

$$P(a < x < b) = \int_a^b f(x)\,dx$$

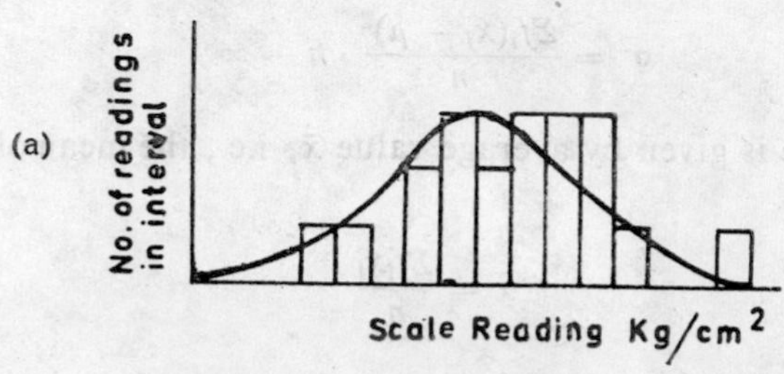

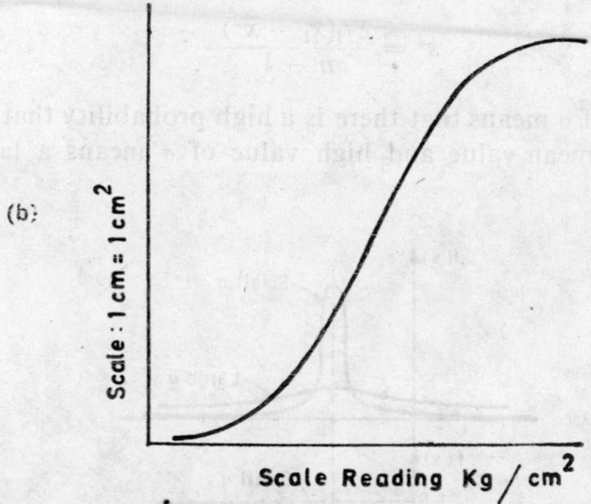

Fig. 3.2 Probability distribution functions

The probability information is sometimes given in terms of the cumulative distribution $F(x)$, which is defined as

$$F(x) \triangleq \text{probability that the measurement is less than any chosen value of } x$$

$$= \int_{-\infty}^{x} f(x)\,dx$$

This is illustrated in Fig. 3.2(b).

As has been said that the measured values are always uncertain and this uncertainty may be due both to systematic and random errors. The random errors follow a statistical distribution and generally it is a normal or Gaussian distribution and is represented by

$$f(x) = \frac{1}{\sqrt{2\pi}\sigma} \exp\{-(x-\mu)^2/2\sigma^2\}$$

where μ is the mean value and σ is the standard deviation. They are defined as

$$\mu = \frac{\Sigma f_i x_i}{n}, n \to \infty$$

and

$$\sigma^2 = \frac{\Sigma f_i (x_i - \mu)^2}{n}, n \to \infty$$

An estimate of μ is given by average value $\bar{x}$, i.e., the mean of the sample defined as

$$\bar{x} = \frac{\Sigma f_i x_i}{n}$$

and the estimate of σ by s, called the sample standard deviation,

$$s^2 = \frac{\Sigma f_i (x_i - \bar{x}^2)}{n - 1}$$

Small values of σ means that there is a high probability that the reading will be near the mean value and high value of σ means a larger scatter (Fig. 3.3).

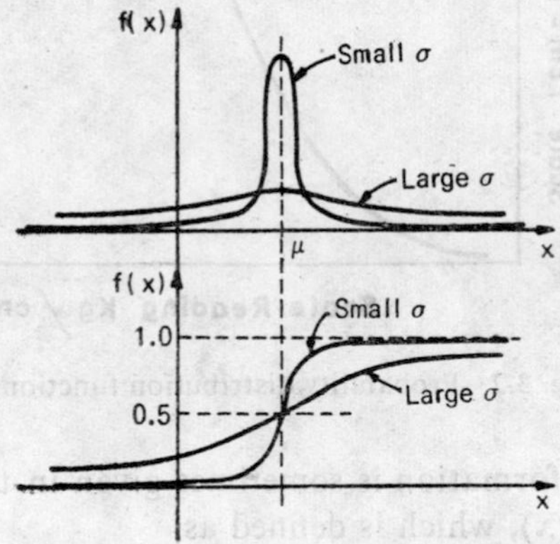

Fig. 3.3 Gaussian distribution for small and large σ values

3.6 Method to calculate $\bar{x}$ and s from the measurements

Using the data of Table 3.1 a procedure for the calculation of $\bar{x}$ and s can be developed. First lump the original measurements into ranges instead of considering each measurement. This is known as *Shepard's correction.* Its effect on the values of $\bar{x}$ and s is negligibly small and may be ignored. Then change the variables x, using numbers from 0 to as high as necessary. This way we introduce an amplification of 400 by changing x variable from a range of 0.0025 per group to a range of 1.0 per group. This is shown in Table 3.2.

TABLE 3.2 Procedure to calculate $\overline{x}$ and s

(Assume mean $\overline{x} = 9$)

S. No.	Original x	New	f_i	$(x_i - \overline{x}')$	$f_i(x_i - \overline{x}')$	$f_i(\overline{x} - \overline{x}')^2$
1.	0.08125	0	1	−9	−9	+81
2.	0.08275	1	1	−8	−8	+64
3.	0.08625	2	0	−7	0	0
4.	0.08875	3	2	−7	−12	+72
5.	0.09125	4	2	−5	−10	+50
6.	0.09375	5	4	−4	−16	+64
7.	0.09625	6	9	−30	−27	+81
8.	0.09875	7	22	−2	−44	+88
9.	0.10125	8	18	−1	−18	+18
10.	0.10375	9	25	0	Sum = −144	0
11.	0.10625	10	23	+1	+23	+23
12.	0.10875	11	15	+2	+30	+60
13.	0.11125	12	8	+3	+21	+63
14.	0.11375	13	12	+4	+48	+192
15.	0.11625	14	8	+5	+40	+200
16.	0.11875	15	8	+6	+46	+288
17.	0.12125	16	2	+7	+14	+98
18.	0.12375	17	2	+8	+16	+128
19.	0.12625	18	1	+9	+9	+81
			162		Sum = +249	Sum=+1651
					Net = +105	

Now make an approximate guess for the mean value, i.e., looking for the position of centroid. Say, in this case it is $x = 9$. This is an apparent mean, denoted by $\overline{x}'$. Using formula

$$\overline{x} = \overline{x}' + \frac{\Sigma f_i(x_i - \overline{x}')}{n}$$

$$= 9 + \frac{105}{162} = 9.65$$

Comparing this with the original strain measurements, it may be seen that it corresponds to a value of strain of $0.10375 + 0.65 \times 0.0025 = 0.10537$. The true mean value is therefore $\overline{x} = 0.10537$.

The calculation of standard deviation is carried out by the following formula

$$\Sigma f_i(x_i - \overline{x})^2 = \Sigma f_i[(x_i - \overline{x}') - (x - \overline{x}')]^2$$
$$= \Sigma f_i(x_i - \overline{x}')^2 - n(\overline{x} - \overline{x}')^2$$

From Table 3.2, we have $\Sigma f_i(x_i - \overline{x}')^2 = 1651$, and $$\bar{x} - \bar{x}' = 0.65.$$

Therefore $$\Sigma f_i(x_i - \bar{x})^2 = 1651 - 162 \times (0.65)^2 = 1583.$$

The standard deviation $s = \sqrt{\dfrac{1583}{162 - 1}} = 3.12.$

Since there is an amplification of 400, the final result for s will be $3.12/400 = 0.0078$.

Therefore the average value of strain = 0.10537,

and standard deviation = 0.0078.

3.7 Confidence limits

It can be shown that

68.27% measurements lie within $\pm 1\sigma$ of $\bar{x}$,

95.45% measurements lie within $\pm 2\sigma$ of $\bar{x}$, and

99.73% measurements lie within $\pm 3\sigma$ of $\bar{x}$.

The total area under the curve includes all the measurements, that means 100%. The value $\pm a\sigma$ is called the confidence limits, where $a = 1, 2, 3$ depends on the chosen reliability. It is common to use a value of $a = 1.96$ which corresponds to a reliability of 95%.

The standard deviation σ is related to a very large number of measurements. Naturally in practice the number of measurements taken is small. Therefore the standard deviation σ must be calculated from the sample standard deviation s. For this purpose conversion Tables 3.3 and 3.4 may be used.

When the mean values are obtained from large number of samples, the distribution of mean values is again Gaussian with mean μ and standard deviation $s/\sqrt{n}$, where n is the number of samples used to obtain mean values.

3.8 Conversion tables

For the measurements

For 68.27% reliability $x \pm 1\sigma = \bar{x} \pm t_1 s$

For 95.00% reliability $x \pm 1.96\sigma = \bar{x} \pm t_2 s$

For 99.73% reliability $x \pm 3\sigma = \bar{x} \pm t_3 s$

TABLE 3.3

n	2	5	10	20	50	100	∞
t_1	1.8	1.15	1.06	1.03	1.01	1.00	1.00
t_2	12.7	2.80	2.30	2.10	2.00	2.00	1.96
t_3	235	6.60	4.10	3.40	3.20	3.10	3.00

For the mean $\overline{x}$

For 68.27% reliability $\overline{x} \pm \frac{1000\sigma}{\sqrt{n}} = \overline{x} \pm \frac{t_1}{\sqrt{n}} s$

For 95.00% reliability $\overline{x} \pm \frac{1.96\sigma}{\sqrt{n}} = \overline{x} \pm \frac{t_3}{\sqrt{n}} s$

For 99.75% reliability $\overline{x} \pm \frac{3\sigma}{\sqrt{n}} = \overline{x} \pm \frac{t_3}{\sqrt{n}} s$

TABLE 3.4

n	2	5	10	20	50	100	∞
$t_1/\sqrt{n}$	1.3	0.51	0.34	0.23	0.14	0.10	0
$t_2/\sqrt{n}$	9.0	1.24	0.72	0.47	0.28	0.20	0
$t_3/\sqrt{n}$	166	3.00	1.29	0.77	0.47	0.31	0

REJECTION OF MEASUREMENTS

It is shown above that 99.73% of all the measurements lie within the confidence limits of $\pm 3\sigma$. Therefore, a measurement lying outside this range may be regarded as having a genuinely extraneous cause such as personal error in measurement, malfunction of apparatus, etc., and may therefore be rejected.

3.9 Testing a distribution for normalcy

The $\pm a\sigma$ bounds put on the best value of measurements are based on the fact that the measurements are assumed to follow the Gaussian distribution. Therefore, it is advisable to test the data whether it follows Gaussian distribution. Two methods are available for this.

(i) Use of probability graph paper, and
(ii) χ^2 test.

If one takes the cumulative distribution function for a Gaussian distribution and suitably distorts the vertical scale of the graph, the curve can be plotted as a straight line. Such graph paper is commercially available and may be used to give a rough qualitative test for the conformity to the Gaussian distribution.

Another method of testing for normalcy involves the use of χ^2 (Chi-square) statistical test. The χ^2 test is applied by comparing the number of times n_0 an event was observed to happen with the number of times n_e that the event would be expected to happen if the hypothesis were true. χ^2 is then defined as

$$\chi^2 = \Sigma \frac{(n_0 - n_e)^2}{n_e}$$

The hypothesis is that the measurements follow Gaussian distribution. The value of n_e can be obtained from the known values of n, $\overline{x}$ and s using statistical tables. The calculation of χ^2 for the earlier example on strain measurement is shown in Table 3.5. The only point that needs consideration is that it is necessary to group together thinly populated regions at both ends of the distribution, so that no group has associated with it a value of n_e less than about five.

TABLE 3.5 Procedure to calculate χ^2

x	n_0	n_e	$n_0 - n_e$	$(n_0 - n_e)^2$	$(n_0 - n_e)^2/n_e$
0–4	6	7.8	1.8	3.24	0.41
5	4	6.8	2.8	7.88	1.16
6	9	10.4	1.4	1.96	0.20
7	21	14.4	6.6	43.6	3.03
8	18	18.0	0.0	0.0	0.00
9	25	20.2	4.8	23.0	1.14
10	23	20.6	2.4	5.8	0.28
11	15	18.9	3.9	15.2	0.81
12	7	15.6	8.6	74.0	4.74
13	12	11.7	0.3	0.09	0.01
14	8	7.9	0.1	0.01	0.00
15	8	4.8	3.2	10.2	2.13
16–18	5	4.5	0.5	0.25	0.05
					Sum = 13.96 = χ^2

In order to apply the criterion, one finds the degrees of freedom. Referring to Table 3.5 there are 13 rows and three constants n, $\bar{x}$ and s which are used to fit the data. Therefore, the degrees of freedom are ten. With the number of degrees of freedom and experimental χ^2, and consulting Table 3.6 one obtains the probability P that this value of χ^2 or higher values, could occur by chance. If $\chi^2 = 0$ the expected and experimental distributions match exactly. However, one must be very cautious in making inference with high values of P. As an example, consider a float level controller used to control the level of water, where the recorder always shows the set point level with no deviations, whatsoever. This is not experimentally true and the experimenter will be suspicious and look for the faults in the recorder.

On the other hand, larger the value of χ^2, the larger the disagreement between expected and experimental distributions, or smaller is the probability that the observed distribution matches the expected distribution.

A good rule of thumb is that if P lies between 0.1 and 0.9, the observed distribution may be considered to follow the expected distribution. For the values of P below 0.02 and higher than 0.98, the expected distribution may be considered unlikely.

It is worthwhile to point out one feature of χ^2 which has general applications. Inspection of Table 3.6 will show that if the hypothesis is true, χ^2 is of the order of magnitude of the number of degrees of freedom.

TABLE 3.6 Chi-square—*P* is the probability that the value in the table will be exceeded for a given number of degrees of freedom *F*

F= \ *P*	0.995	.990	0.975	0.0950	0.900	0.750	0.500	0.250	0.100	0.050	0.025	0.010	0.0
1	$0.0^{4}393$	$0.0^{3}157$	$0.0^{3}982$	$0.0^{2}393$	0.0158	0.102	0.455	1.32	2.71	3.84	5.02	6.63	7.8
2	0.0100	0.0201	0.0506	0.103	0.211	0.575	1.39	2.77	4.61	5.99	7.38	9.21	10.6
3	0.0717	0.115	0.216	0.352	0.584	1.21	2.37	4.11	6.25	7.81	9.35	11.1	12.8
4	0.207	0.297	0.484	0.711	1.06	1.92	3.36	5.39	7.78	9.49	11.1	13.3	14.9
5	0.412	0.554	0.831	1.15	1.61	2.67	4.35	6.63	9.24	11.1	12.8	15.1	16.7
6	0.676	0.872	1.24	1.64	2.20	3.45	5.35	7.84	10.6	12.6	14.4	16.8	18.5
7	0.989	1.24	1.69	2.17	2.83	4.25	6.35	9.04	12.0	14.1	16.0	18.5	20.3
8	1.34	1.65	2.18	2.73	3.49	5.07	7.34	10.2	13.4	15.5	17.5	20.1	22.0
9	1.73	2.09	2.70	3.33	4.17	5.90	8.34	11.4	14.7	16.9	19.0	21.7	23.6
10	2.16	2.56	3.25	3.94	4.87	6.74	9.34	12.5	16.0	18.3	20.5	23.2	25.2
11	2.60	3.05	3.82	4.57	5.58	7.58	10.3	13.7	17.3	19.7	21.9	24.7	26.8
12	3.07	3.57	4.40	5.23	6.30	8.44	11.3	14.8	18.5	21.0	23.3	26.2	28.3
13	3.57	4.11	5.01	5.89	7.04	9.30	12.3	16.0	19.8	22.4	24.7	27.7	29.8
14	4.07	4.66	5.63	6.57	7.79	10.2	13.3	17.1	21.1	23.7	26.1	29.1	31.3
15	4.60	5.23	6.26	7.26	8.55	11.0	14.3	18.2	22.3	25.0	27.5	30.6	32.8
16	5.14	5.81	6.91	7.96	9.31	11.9	15.3	19.4	23.5	26.3	28.8	32.0	34.3
17	5.70	6.41	7.56	8.67	10.1	12.8	16.3	20.5	24.8	27.6	30.2	33.4	35.7
18	6.26	7.01	8.23	9.39	10.9	13.7	17.3	21.6	26.0	28.9	31.5	34.8	37.2
19	6.84	7.63	8.91	10.1	11.7	14.6	18 3	22.7	27.2	30.1	32.9	36.2	38.6
20	7.43	8.26	9.59	10.9	21.4	15.5	19.3	23.8	28.4	31.4	34.2	37.6	40.0
21	8.03	8.90	10.3	11.6	13.2	16.3	20.3	24.9	29.6	32.7	35.5	38.9	41.4
22	8.64	9.54	11.0	12.3	14.0	17.2	21.3	26.0	30.8	33.9	36.8	40.3	42.8
23	9.26	10.2	11.7	13.1	14.8	18.1	22.3	27.1	32.0	35.2	38.1	41.6	44.2
24	9.89	10.9	12.4	13.8	15.7	19.0	23.3	28.2	33.2	36.4	39.4	43.0	45.6
25	10.5	11.5	13.1	14.6	16.5	19.9	24.3	29.3	34.4	37.7	40.6	44.3	46.9
26	11.2	12.2	13.8	15.4	17.3	20.8	25.3	30.4	35.6	38.9	41.9	45.6	48.3
27	11.8	12.9	14.6	16.2	18.1	21.7	26.3	31.5	36.7	40.1	43.2	47.0	49.6
28	12.5	13.6	15.3	16.9	18.9	22.7	27.3	32.6	37.9	41.3	44.5	48.3	51.0
29	13.1	14.3	16.0	17.7	19.8	23.6	28.3	33.7	39.1	42.6	45.7	49.6	52.3
30	13.8	15.0	16.8	18.5	20.6	24.5	29.3	34.8	40.3	43.8	47.0	50.9	53.7

Therefore,

$$\frac{(n_0 - n_e)^2}{n_e} \simeq 1$$

or

$$n_0 \simeq n_e \pm n_e^{1/2}$$

The interpretation of this result is that whenever the value n_e is to be obtained it is likely that $n_e \pm \sqrt{n_e}$ is observed. For example, if a coin is tossed 1000 times, one expects 500 heads, but in practice it is likely to be 'off' by a number of the order of magnitude of $\sqrt{500}$ or 22. As the number of measurements increases, total error increases by $n_e^{1/2}$, while the fractional error decreases by $n_e^{-1/2}$.

However, more important aspects of statistical analysis of measurements which is often used for calibration will be discussed now. For most of the instruments, but not all, the input-output relation is ideally linear. In some of the instruments, the operation is very much restricted over a linear region. The average calibration curve for such an instrument is generally taken as a straight line which fits the scattered data points best. The most common criterion for best line fit is the least square method, which minimises the sum of the squares of the vertical deviations of the data point from the fitted line. The equation of straight line is taken as

$$q_0 = mq_i + a$$

where q_0 is the output quantity, q_i is the input quantity, m, the slope of the straight line, and a, a constant.

The values of constants m and b are so chosen that the sum of the square of deviations is minimum, i.e., minimise the quantity S as

$$S = \sum^{n} \{q_0 - (mq_i + a)\}^2$$

Thus

$$m = \frac{n\Sigma q_i q_0 - (\Sigma q_i)(\Sigma q_0)}{n\Sigma q_i^2 - (\Sigma q_i)^2}$$

and

$$a = \frac{(\Sigma q_0)(\Sigma q_i^2) - (\Sigma q_i q_0)(\Sigma q_i)}{n\Sigma q_i^2 - (\Sigma q_i)^2}$$

where n is the total number of data points.

Assuming that q_i were fixed, and repeated measurements were made on q_0, the latter will show scatter for a fixed value of q_i. Therefore, m and a will also exhibit scatter with their standard deviations given by the following expressions:

$$s_m^2 = \frac{ns_{q_0}^2}{n\Sigma q_i^2 - (\Sigma q_i)^2}$$

$$s_a^2 = \frac{s_{q_0}\Sigma q_i^2}{n\Sigma q_i^2 - (\Sigma q_i)^2}$$

where $s_{q_0}^2 = \dfrac{\Sigma\{q_0 - (mq_i + a)\}^2}{n}$ is the variance of q_0.

Assuming Gaussian distribution and 95% reliability limits, one can set the bounds on both m and a as

$$m = m \pm 1.96 s_m$$

and
$$a = a \pm 1.96 s_a$$

This procedure is used to obtain the calibration curve where observations at constant input are made and a best line fit is obtained. However, this calibration curve is later used to obtain the value of input quantity from the output value. It is, therefore, necessary to put bounds on the input quantity which is being measured. From the least square line

$$q_i = \frac{q_0 - a}{m}$$

The q_i values computed this way must have some $\pm$ error limits. The standard deviation s_{q_i} of q_i is computed from

$$s_{q_i} = \frac{1}{n} \Sigma \left(\frac{q_0 - b}{m} - q_i \right)^2 = \frac{s_{q_0}^2}{m^2}$$

Thus the 95% reliability limits of q_i are $q_i \pm 1.96(s_{q_0}/m)$. Another common method of stating bounds on the error uses the probable error, e_p. This is defined as

$$e_p = 0.6740 s$$

A range of $\pm e_p$ corresponds to 50% reliability.

It must be noted that if s_{q_0} is assumed to be same for any value of q_i, a set of data q_0 need not be repeated for same value of q_i, but the standard deviation s_{q_0} can be obtained from the set of data where one output q_0 is measured for any input q_i.

These concepts are explained with the help of the following example.

Example: From the data given below, obtain a line of best fit using the method of least square.

q_i lps	q_0 lps
0.0	−1.12
1.0	0.21
2.0	1.18
3.0	2.09
4.0	3.33
5.0	4.50
6.0	5.26
7.0	6.59
8.0	7.73
9.0	8.68
10.0	9.80

Solution

q_i	q_0	q_i^2	q_0q_i	$q_{0(cal)}$	$\lvert q_0 - q_{0(cal)}\rvert$	$(q_0 - q_{0(cal)})^2$
0.0	−1.12	0.0	−0.00	−1.03	0.09	0.009
1.0	0.21	1.0	0.21	0.06	0.15	0.023
2.0	1.18	4.0	2.36	1.14	0.04	0.002
3.0	2.09	9.0	6.27	2.22	0.13	0.017
4.0	3.33	16.0	13.32	3.30	0.03	0.001
5.0	4.50	25.0	22.50	4.39	0.11	0.013
6.0	5.26	36.0	31.56	5.47	0.21	0.044
7.0	6.59	49.0	46.13	6.55	0.04	0.002
8.0	7.73	64.0	61.84	7.63	0.10	0.009
9.0	8.68	81.0	78.12	8.72	0.04	0.002
10.0	9.80	100.0	98.00	9.79	0.00	0.000
55.0	48.25	385.0	360.31	48.24	0.94	0.122

Thus
$$m = \frac{n\Sigma q_i q_0 - (\Sigma q_i)(\Sigma q_0)}{n\Sigma q_i^2 - (\Sigma q_i)^2}$$

$$= \frac{11\times 360.31 - 55\times 48.25}{11\times 385 - 55\times 55} = 1.08$$

and
$$a = \frac{(\Sigma q_0)(\Sigma q_i^2) - (\Sigma q_i q_0)(\Sigma q_i)}{n\Sigma q_i^2 - (\Sigma q_i)^2}$$

$$= \frac{48.25\times 385 - 360.31\times 55}{1210.0} = -1.026$$

Thus $\quad q_0 = 1.08\times q_i - 1.026$

Now make calculations to set bounds on m and a as follows:

$$s_{q_0}^2 = \frac{0.122}{11} = 0.011$$

Thus
$$s_{q_0} = 0.10$$

Now
$$s_m^2 = \frac{0.122}{1210.0} = 0.10\times 10^{-3}$$

or
$$s_m = 1.0\times 10^{-2}$$

Similarly,

$$s_a^2 = \frac{0.011\times 385}{1210.0} = 3.5\times 10^{-3}$$

$$s_a = 5.92\times 10^{-2}$$

Result:
$$m = 1.08$$
$$s_m = 1.0\times 10^{-2}$$
$$a = -1.026 \text{ lps}$$

$$s_b = 5.92 \times 10^{-2} \text{ lps}$$

3.10 Propagation of error

The nomenclature followed here is:

reading ± uncertainty = result

The term 'reading' should be called 'corrected reading' after applying the correction. The correction has the same magnitude as the systematic error, but opposite sign. It is, however, a usual practice to call the 'corrected reading' as a 'reading' only.

In many experiments, results from different measuring instruments are used to compute the value of a particular physical quantity. The following two questions need consideration:

(i) if the uncertainty of measurement of each instrument is known, what is the uncertainty of the computed result?

(ii) if a certain uncertainty in the computed result is desired, what uncertainty is allowed in the individual instruments?

Consider a problem of computing a quantity, g, where g is a known function of n independent variables $u_1, u_2, \ldots, u_n$. That is,

$$g = f(u_1, u_2, \ldots, u_n)$$

These variables $(u_1, \ldots, u_n)$ are measured quantities and different measuring instruments are used to measure them. Let these values be uncertain by $\pm\Delta u_1$, $\pm\Delta u_2, \ldots \pm\Delta u_n$. These uncertainties will cause an uncertainty of magnitude Δg in the value of g. The method of computation of Δg depends on the following two cases:

(i) when $\pm\Delta u_n$ is an absolute limit on the uncertainties as is the case with systematic errors, and

(ii) when $\pm\Delta u_n$ is a statistical bound such as ± 1.96 or e_p limits as is the case with systematic errors of unknown sign or accidental errors.

CASE 1: $g = f(u_1, u_2, \ldots, u_n)$

Therefore

$$g \pm \Delta g = f(u_1 \pm \Delta u_1, u_2 \pm \Delta u_2, \ldots u_n \pm \Delta u_n)$$

If $\Delta u_n \ll u_n$, expanding by Taylor's series

$$g \pm \Delta g = f(u_1, u_2, \ldots, u_n) \pm \Delta u_1 \frac{\partial f}{\partial u_1} \pm \Delta u_2 \frac{\partial f}{\partial u_2} \pm \ldots \pm \Delta u_n \frac{\partial f}{\partial u_n}$$
$$\pm \text{ higher order terms}$$

Therefore, the absolute error E_a is

$$E_a = \Delta g = \left|\Delta u_1 \frac{\partial f}{\partial u_1}\right| + \left|\Delta u_2 \frac{\partial f}{\delta u_2}\right| + \ldots + \left|\Delta u_n \frac{\partial f}{\partial u_n}\right|$$

where absolute values have been used; the total error is, therefore, obtained by adding the numerical values of individual errors. This form of equation is very useful since it shows which variable (u_n) exerts the strongest influence on the uncertainty of overall result.

The relative percentage error E_r is given by

$$E_r = \frac{\Delta g}{g} \times 100 = 100 \frac{E_a}{g}$$

and the computed results may be expressed as either $g \pm E_a$ or $g + E_r\%$, and the interpretation is that this error will not be exceeded in the measurements since this is the way Δu_n are defined.

When the second question is considered, i.e., where a certain overall uncertainty in the result is desired and the uncertainty of the components is to be calculated, the problem becomes extremely complicated. It is simplified when the 'method of equal effects' is used, i.e., it is assumed that each measurement is being made such that

$$\left|\Delta u_1 \frac{\partial f}{\partial u_1}\right| = \left|\Delta u_2 \frac{\partial f}{\partial u_2}\right| + \ldots = \left|\Delta u_n \frac{\partial f}{\partial u_n}\right|$$

or

$$n \frac{\partial f}{\partial u_i} \Delta u_i = \Delta g \ (i = 1, 2, \ldots, n)$$

and hence

$$\Delta u_i = \frac{\Delta g}{n\left(\dfrac{\partial f}{\partial u_i}\right)}$$

In actual experimentation, if a particular Δu_i turns out smaller than what possibly can be achieved by the instrument available, it is possible to relax this requirement if some other Δu_n can be made smaller than the value given by the equation. That is to say, some instruments can give better accuracy than demanded by the above equation, while others may be unable to meet the requirement. In such cases, it may still be possible to meet overall uncertainty requirements.

CASE 2: When Δu_n is considered as statistical bound such as $\pm 1.96s$ limits, i.e., Δu_n follows a normal distribution, then the computed overall uncertainty would also follow a Gaussian distribution. Thus in this, the proper method of combining such uncertainties is according to root-sum-square formula, i.e.

$$E_{a_{rss}} = \left[\left(\Delta u_1 \frac{\partial f}{\partial u_i}\right)^2 + \left(\Delta u_2 \frac{\partial f}{\partial u_2}\right)^2 + \ldots + \left(\Delta u_n \frac{\partial f}{\partial u_n}\right)^2\right]^{1/2}$$

and a 99.73% reliability on g can be expressed by

$$g \pm 3E_{a_{rss}}$$

This equation always gives a smaller value of error than given by the earlier equation.

Further, when a known uncertainty of the computed result is desired, again using the 'method of equal effects',

$$\Delta u_i = \frac{\Delta g}{\sqrt{n}\left(\frac{\partial f}{\partial u_i}\right)}$$

These statistical concepts developed thus far should be implemented in regular experimentation.

Exercises

1. A resistance arrangement of 100 ohms is desired. Two resistances of 200 $\pm$ 0.1 ohms and two resistances of 50.0 $\pm$ 0.02 ohms are available. Which should be used: a series arrangement with 50 ohms resistors or a parallel arrangement with the 200 ohms resistors? Calculate the uncertainty for each arrangement.
2. In a strain gauge experiment, the following equation is used to calculate strain ϵ:
$$\epsilon = \frac{e_0}{E_b}\cdot\frac{1}{F}\cdot\frac{(R_a + R_g)^2}{R_a R_g}$$
where the output voltage e_0, excitation voltage E_b, gauge resistance R_g, gauge factor F and resistance R_a are measured to $\pm 2\%$. Calculate the possible uncertainty in the computed value ϵ.
3. The discharge coefficient C_q of an orifice can be found by collecting the water that flows through it during a time interval when it is under a constant head h. The formula is
$$C_q = \frac{W}{t\rho A\sqrt{2gh}}$$
Find C_q with its uncertainty if
$W = 400 \pm 0.2$ kg, $A = \pi d^2/4$, $d = 15 \pm 0.02$ mm, $t = 500.00 \pm 2$ sec,
$g = 9.81 \pm 0.1\%$ m/sec², $\rho = 10^3 \pm 0.1\%$ kg/m³, $h = 4 \pm 0.04$ m.
Consider both the cases
(i) the uncertainties are absolute values, and
(ii) the uncertainties are $\pm 3s$ statistical limits.
4. The expected relationship between q_0 and q_i is of the form
$$q_0 = bq_i^a$$
Plot the following data and also obtain a best fit using the least square method:

q_0	q_i
13.0	1720
15.0	1350
17.0	1600
18.0	1000
19.2	1240
21.5	2150
28.7	2470
32.1	2600
33.0	2030
42.7	2980
44.0	3400
52.0	3000
60.0	3300
80.0	2100
130.0	3900

5. The following set of data points that q_i is expected to be a quadratic function of q_0. Obtain the quadratic function by the method of least squares.

q_i	q_0
2.1	1
8.9	2
19.3	3
40.0	4
115.0	5
152.0	6

6. Devise a method to plot Gaussian distribution and its cumulative distribution such that a straight line will result.

4

Measurement of Force and Torque

4.1 Introduction

Force is represented mathematically as a vector with a point of application. Therefore force measurement involves the determination of magnitude as well as the direction of force. The measurement of force may be carried out by means of the following:

(i) *Direct methods*: By direct comparison with a known gravitational force on a standard mass (say, by a balance).

(ii) *Indirect methods*: By measuring the effect of force on a body, for example, by

(a) measuring the acceleration of a body of known mass to which the force is applied.

(b) applying the force to an elastic member and measuring the resultant effect.

4.2 Direct methods

The weight of a body is the force exerted on the body by the earth's gravitational attraction. Therefore

$$F = mg$$

The unknown force may be balanced against the gravitational force (mg) on the standard mass m. The value of m and g should be accurately known in order to know the magnitude of the gravitational force. The mass is a fundamental quantity and its standard, kilogram, is kept in a vault at Sevres, France. The other masses can be compared with this standard with a precision of a few parts in 10^9. On the other hand, g is a derived quantity but still makes a convenient standard. This can be measured with an accuracy of 1 part in 10^6.

In order to understand the principle underlying this method consider an analytical balance as shown in Fig. 4.1. The balance arm rotates about the knife edge at point O and is shown in an unbalanced position as indicated by an angle ϕ. Point G represents the centre of gravity of the arm, and d_G is the distance between these points. W_B is the weight of the balance arm and the pointer acting at GW_1 is the unknown force and W_2 is the force due to standard mass. In the balanced position $W_1 = W_2$, and

ϕ is zero. Therefore, the weight of the balance arm and pointer do not influence the measurements.

In order to understand the working of this kind of balance, a little on the sensitivity of this balance will now be dealt with. The sensitivity may be defined by

$$S = \frac{\phi}{W_1 - W_2} = \frac{\phi}{\Delta W}$$

It is, therefore, a measure of the angular deflection per unit unbalance between the two weights W_1 and W_2. Referring to Fig. 4.1, for equilibrium we have

$$W_1(L \cos \phi - d_B \sin \phi) = W_2(L \cos \phi + d_B \sin \phi) + W_B d_G \sin \phi.$$

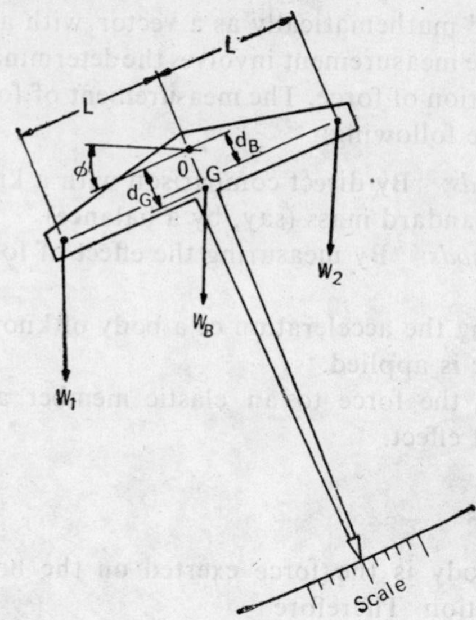

Fig. 4.1 Schematic of an analytical balance

Assuming very small angular deflection, the above equation is rewritten as

$$W_1(L - d_B\phi) = W_2(L + d_B\phi) + W_B \cdot d_G \cdot \phi$$

or

$$\frac{\phi}{W_1 - W_2} = \frac{L}{(W_1 + W_2)d_B + W_B d_G}$$

At a near equilibrium condition, $W_1 \simeq W_2 = W$ and hence

$$S = \frac{\phi}{\Delta W} = \frac{L}{2W d_B + W_B d_G}$$

If the balance arm is constructed so that $d_B = 0$, then

$$S = \frac{L}{W_B d_G}$$

The sensitivity is independent of the load. This is, therefore, a very important result. Precision balances are available which have an accuracy of 1 part in 10^8.

While comparing force with a gravitational force on a standard mass using balance, sufficient care should be exercised to correct the result for buoyancy forces acting on the mass. The range of measurement can be considerably extended by using levers as is done in weighing platforms.

4.3 Indirect methods

The measurement of force by measuring acceleration of a standard mass when force is acting on this is based on the principle that

$$F = ma, \text{ where } a \text{ is the acceleration.}$$

The measurement of a can be carried out by accelerometers. However, this method is of limited application since the force determined is the resultant force acting on the mass. Often, several unknown forces are acting and they cannot be separately measured by this method. This method is represented schematically in Fig. 4.2.

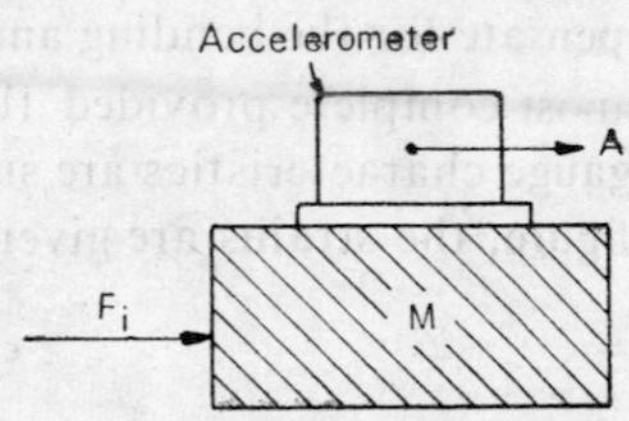

Fig. 4.2 Force measurement by accelerometer

Elastic elements are frequently used for the measurement of force because of their large range, continuous monitoring, ease of operation and ruggedness. They furnish an indication of the magnitude of force through displacement measurement. They are used both for dynamic and static force measurements.

(i) Spring

A simple spring is an example of a force displacement transducer. If the displacement from the equilibrium position is y, then the force F is given by (within a certain limit)

$$F = Ky$$

where K, force per unit displacement, is the spring constant.

The displacement bears a linear relation with the applied force. This system is quite satisfactory for steady state measurements.

(ii) Axially Loaded Members

In this method the members (say, a bar or rod) are axially loaded either in compression or tension. The displacement y of the free end, when a force F is applied, is given by

$$y = \frac{F \cdot L}{A \cdot E}$$

where A is the area, L = length, and E = Young's modulus of the material.

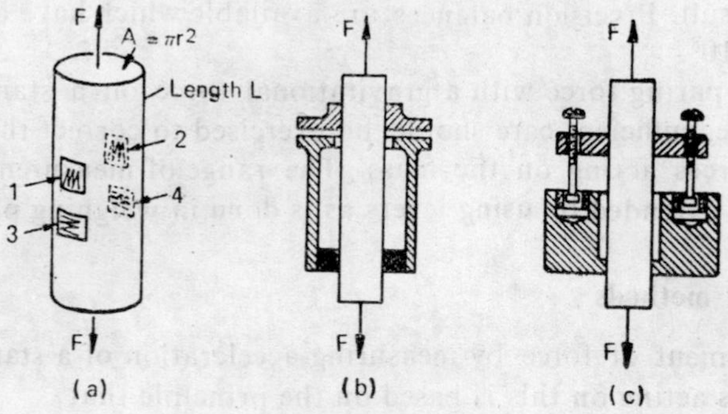

Fig. 4.3 Axially loaded members for force measurement

Fig. 4.3 (a) shows a rod subjected to force F and where the strains (y/L) is measured by strain gauges. The strain gauges are so mounted as to compensate for the bending and temperature variations. The compensation is almost complete provided the gauges are symmetrically mounted and the gauge characteristics are sufficiently uniform. For the gauges shown in the figure, the strains are given by

$$\epsilon_1, \epsilon_2 = \frac{F}{AE}$$

and

$$\epsilon_3, \epsilon_4 = \frac{-\mu F}{AE}$$

where μ is the Poisson's ratio of the material. The gauges 1 and 2 measure the axial strain, while 3 and 4 measure the lateral strain. The spring constant of the member may be represented by

$$K = \frac{AE}{L}$$

Instead of measuring displacement through strain gauges, the design of a system could be such as to induce a change in the capacitance or inductance which may be easily measured. This is shown in Figs. 4.3 (b) and (c). The inductance gauges are linear and obviously will not be affected by bending. On the other hand, capacitance gauges are non-linear and will be effected by bending to a certain extent.

(iii) CANTILEVER BEAMS

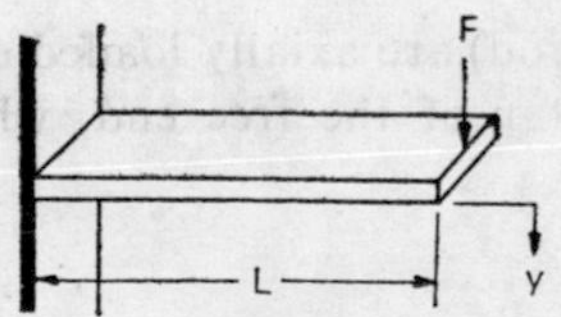

Fig. 4.4 Cantilever elastic element

Cantilevers are ideally suited for the measurement of bending moments about two axes mutually perpendicular to the axis of the beam. If the point of application of load along the beam is known, the bending moment can be directly translated into force. Fig. 4.4 represents a cantilever, where a force F is applied at a distance L from the fixed end. The deflection of the free end is related to the force by the following equation

$$F = \frac{3EIy}{L^3}$$

where I is the moment of inertia of the cross section of the beam. Assuming a cantilever of rectangular cross section of width b and thickness h, the deflection y is

$$y = \frac{4FL^3}{Ebh^3}$$

The spring constant of the system is therefore given by

$$K = \frac{Ebh^3}{4L^3}$$

The displacement can be measured by strain gauges as shown in Fig. 4.5. This arrangement is used to measure the forces P and F applied perpendicular to each other. If the gauges are symmetrically located with respect to the neutral axis of the beam and wired as shown in Fig. 4.5, the two perpendicular forces or force components can be independently measured.

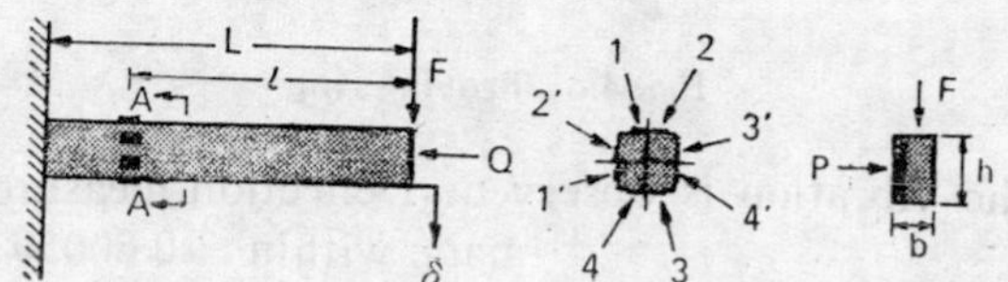

Fig. 4.5 Cantilever elastic element for measuring force

The strains are given by

$$\epsilon_1, \epsilon_2, \epsilon_3, \epsilon_4 = \frac{6Fl}{Ebh^2}$$

where l is the distance from the centre of the gauges to the point of application of force. Similar equations hold good for the other component of force. It may be noted that an axial force component Q will not affect the measurements provided it acts at the centre of cross section.

(iv) RINGS

One of the very useful and important devices under this heading is a *proving ring*. This has been the standard for calibrating tensile-testing machines and is, in general, the means whereby accurate measurement of large static loads may be made. Proving ring can be used over a wide range of loads starting from 1500 to 15×10^5 N.

Fig. 4.6 shows a compression-type proving ring. The compressive load deforms the ring; a sensitive micrometer is employed for the deflection measurement. Bosses are provided to clamp the ring rigidly to avoid rotation. To obtain a precise measurement, one edge of the micrometer is mounted on a vibrating reed device which is plucked to obtain a vibratory motion. The micrometer contact is moved forward until a noticeable

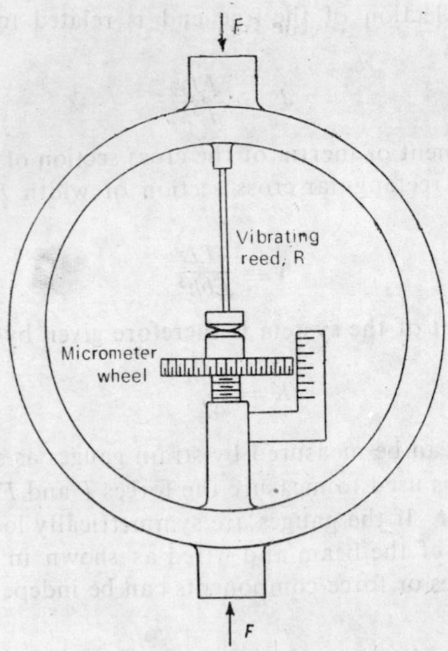

Fig. 4.6 Proving ring

damping of the vibration is observed. Deflection measurements may be made within ± 0.00050 mm with this method.

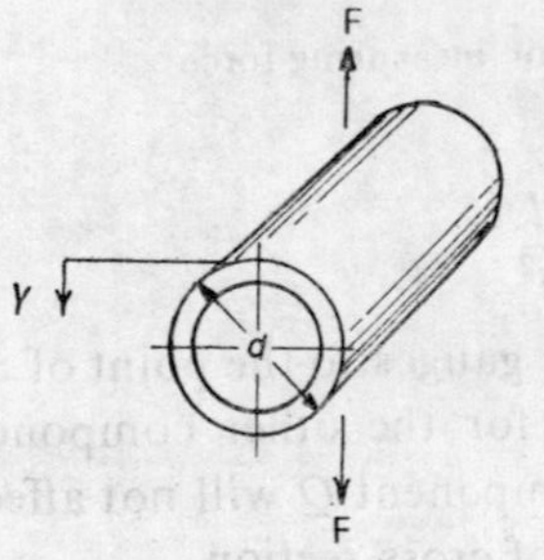

In another variant of proving ring, a differential transformer is used for the measurement of deflection.

Another elastic device which is frequently used for force measurement is a thin ring shown in Fig. 4.7. The force F and deflection y relation for a thin ring is

Fig. 4.7 Thin-ring elastic element

$$F = \frac{16}{\pi/2 - 4/\pi}\left(\frac{EI}{d^3}y\right)$$

where d is the outside ring diameter and I is the moment of inertia about the centroidal axis of the bending section. This equation is derived under the assumption that the thickness of the ring is small compared with the radius. The force is applied perpendicular to the axis of the ring as shown in Fig. 4.7.

Other types of rings used for force measurement are octagonal ring and extended octagonal ring. They are designed so as to have no rotation of the top surface.

(v) Load Cells Using Strain Gauges

Force transducers intended for weighing purposes are called load cells.

Instead of using total deflection as a measure of load, strain-gauge load cells measure load in terms of unit strains. For very large loads, direct tensile-compressive member may be selected. For small loads, strain amplification provided by bending may be employed to advantage. One such tensile compressive cell using four strain gauges is shown in Fig. 4.3 (a). The gauges are so mounted as to give maximum output, and compensation for bending and temperature variations. The sensitivity is $2(1 + \mu)$ times that achieved with a single active gauge in the bridge. Another version of this kind of load cell is given in Fig. 4.8. The basic arrangement is similar for both tensile or compressive load measurements, only the fixtures differ. Compression cells of this kind have been used with a capacity of 15×10^6 N.

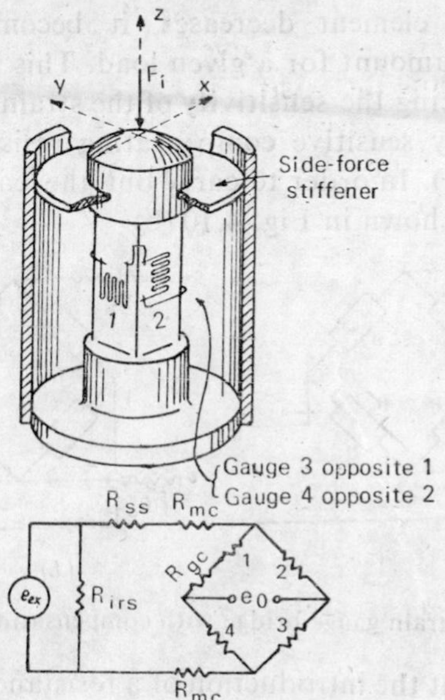

Fig. 4.8 Strain gauge load cell

Figures 4.9 (a) and (b) illustrate proving ring strain gauge load cells. In Fig. 4.9 (a) the bridge output is a function of the bending strains only;

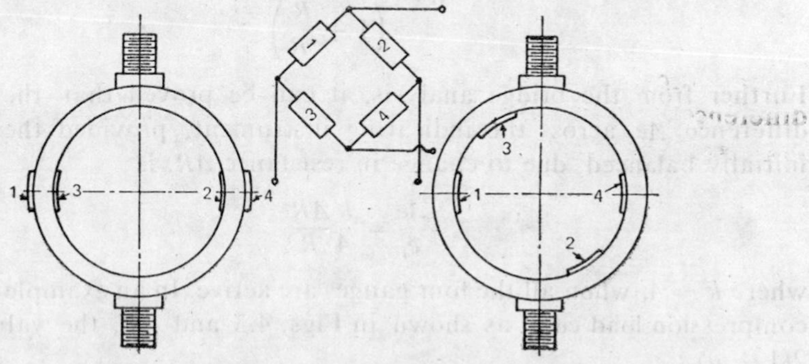

Fig. 4.9 Strain gauge load cells

the axial components being cancelled in the bridge arrangement. The arrangement of Fig. 4.9 (b) provides a somewhat higher sensitivity because the output includes both the bending and axial components sensed by gauges 1 and 4.

So far relatively ideal situations have been considered where the compensation both for bending and temperature variations was achieved by a particular arrangement of gauges. However, careful considerations show that the temperature influences measurements in two ways:

(i) change of dimensions, and
(ii) change in the Young's modulus of the material.

Of these, the latter is more important. Therefore some means of compensation must be employed. As temperature increases, the deflection constant for the elastic element decreases, it becomes more springy and deflects by a greater amount for a given load. This increase in the sensitivity is offset by reducing the sensitivity of the strain gauge bridge through the use of a thermally sensitive compensating resistance element, R_s, as shown in Fig. 4.10 (a). In order to carry out the calibration, usually two resistors are used as shown in Fig. 4.10 (b).

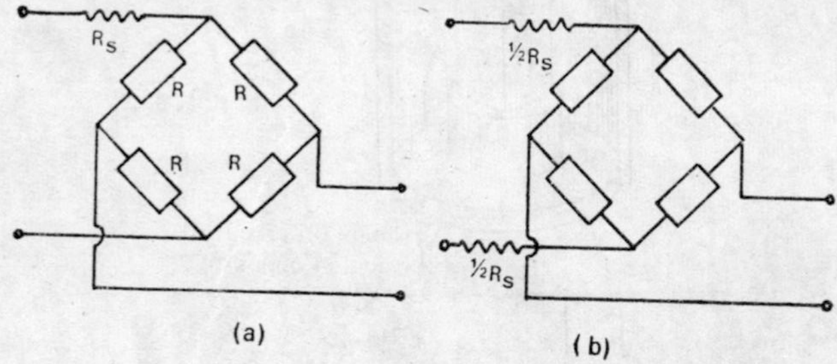

Fig. 4.10 Strain gauge bridge with compensation resistance

It can be shown that the introduction of a resistance, R_s, in an input lead reduces the sensitivity of an equal-arm bridge, each arm of resistance, R, by a bridge factor, n, given as follows:

$$n = \frac{1}{\left(1 + \frac{R_s}{R}\right)}$$

Further from the bridge analysis, it can be proved that the potential difference Δe across the indicating instrument, provided the bridge is initially balanced, due to change in resistance ΔR, is

$$\frac{\Delta e}{e_i} = \frac{k}{4} \frac{\Delta R}{R}$$

where $k = 4$, when all the four gauges are active. In an example of tensile-compression load cells, as shown in Figs. 4.3 and 4.8, the value of k is $2(1 + \mu)$.

When the effect of bridge factor due to resistance R_s is taken into account, this equation becomes

$$\frac{\Delta e}{e_i} = \frac{k}{4} \frac{\Delta R}{R(1 + R_s/R)}$$

On the other hand, the change in the resistance of a strain gauge is related to strain ϵ and gauge factor F_g as

$$\frac{\Delta R}{R} = \epsilon F_g$$

Further the strain is related to Young's modulus, E, and force, F, as $E = \frac{F/A}{\epsilon}$, where A is the area of cross section. Combining these equations, bridge sensitivity may be written as

$$\frac{\Delta e}{F} = \frac{k}{4} \frac{R}{R + R_s} \frac{F_g \cdot e_i}{A \cdot E}$$

Assuming that the compensation for the change in resistance R of a strain gauge due to temperature change is accomplished by proper arrangement, and gauge factor F_g and area A do not vary appreciably with temperature change, then the complete compensation will be achieved provided

$$E(R + R_s) = E(1 + c\Delta T)[R + R_s(1 + b\Delta T)]$$

is independent of temperature change.

This may be achieved when

$$\frac{R_s}{R} = -\frac{c}{b + c}$$

This indicates that temperature compensation may possibly be accomplished through proper balancing of the temperature coefficient of Young's modulus, c, and electrical resistivity, b. Because c is usually negative and because the resistances cannot be negative, it follows that

$$b + c > 0$$

In addition to this,

$$R_s = \rho \frac{L}{A} = -R\left(\frac{c}{b + c}\right)$$

or

$$L = -\frac{RA}{\rho}\left(\frac{c}{b + c}\right)$$

From these relations, specific requirements for compensation may be derived. After a resistance material is selected usually in the form of a wire, the required length may be determined from this equation.

4.4 Torque measurement

The measurement of torque is necessitated in order to obtain load information necessary for stress or deflection analysis. However, torque measurement is often associated with the determination of mechanical power; either power required to operate a machine or power developed by the machine. In this connection, torque measuring devices are commonly known as dynamometers.

There are basically three types of dynamometers:

(i) *Absorption dynamometers*

They absorb the mechanical energy as torque is measured, and hence are particularly useful for measuring power or torque developed by power sources such as engines and motors.

(ii) *Driving dynamometers*

These dynamometers measure torque or power as well as supply energy to operate the devices to be tested. They are particularly useful in determining performance characteristics of devices like pumps and compressors.

(iii) *Transmission dynamometers*

These are passive devices placed at an appropriate location for the purpose of sensing torque at that location. They are sometimes referred to as torque meters. (Driving dynamometers are sometimes called as transmission dynamometers). The first two types can be grouped as mechanical and electrical dynamometers and will be discussed as such here. However, the transmission type is treated separately.

4.5 Mechanical dynamometers

These dynamometers are of absorption type. One of the most familiar and simple devices is the *Prony brake* as shown in Fig. 4.11. The mechanical

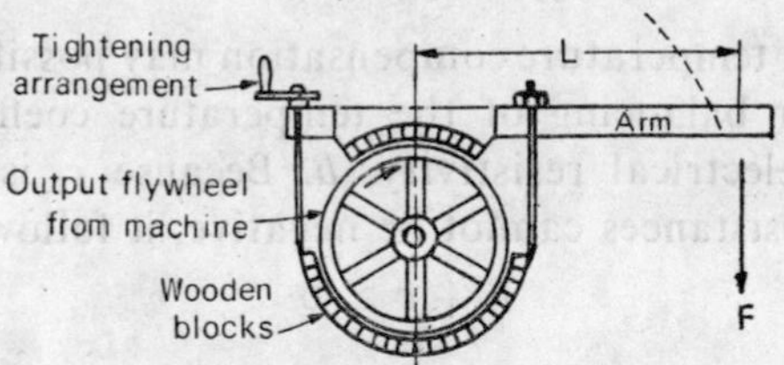

Fig. 4.11 Schematic of a Prony brake

energy is converted to heat through dry friction. The torque exerted on the Prony brake is

$$T = FL$$

where force F is measured by conventional force measuring instruments, say balances or load cells, etc. The power dissipated in brake is calculated from

$$P = 2\pi TN$$

$$= \frac{2\pi FrN}{60}$$

where force F (measured at radius r) is in Newtons, r is the length of reaction arm in metres, N is the angular speed in revolutions per minute, and P = power in watts.

The Prony brake is inexpensive, but it is difficult to adjust and maintain a specific load.

Various other types of brakes are employed for power measurements on mechanical equipment. The water brake, for example, dissipates the output energy through fluid friction.

4.6 Electrical dynamometers

They are of both absorption and driving type. Almost any form of rotating electric machine can be used as a dynamometer, but those specially designed for the purpose are convenient to use. The four possibilities are:

(i) eddy-current dynamometers,
(ii) d.c. dynamometers or generators,
(iii) d.c. motors and generators,
(iv) a.c. motors and generators.

(i) EDDY-CURRENT DYNAMOMETERS. These are strictly of absorption type. The eddy-current dynamometer consists of one (or more) metal disc which is rotated in a magnetic field. The field is produced by field elements or coils excited by an external source and attached to the dynamometer housing, which is mounted in trunnion bearings. As the disc rotates, eddy-currents are generated, and the reaction with the magnetic field tends to rotate complete housing in the trunnion bearings. A schematic form is shown in Fig. 4.12. Torque is measured in the same manner as for Prony or water

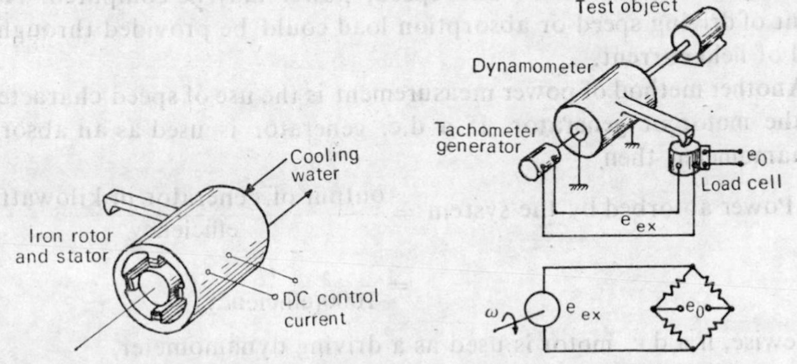

Fig. 4.12 Eddy-current dynamometer Fig. 4.13 d.c. dynamometer

brakes and same equations apply for the torque and power calculations. The eddy-current brake is easily controllable by varying field current, but it cannot produce any torque at zero speed and only small torque at low speeds. The power absorbed is carried away by cooling water circulated through the air gap between rotor and stator. Particular advantages of this type of dynamometers are comparatively small size for a given capacity and characteristics permitting good control at slow speeds.

(ii) d.c. DYNAMOMETERS. The most versatile and accurate dynamometer is the d.c. dynamometer. Here a d.c. machine is mounted in low friction trunnion bearings as shown in Fig. 4.13. This machine can be coupled to either power-absorbing or power-generating devices since it may be connected as either a motor or a generator. When used as an absorption dynamometer, it performs as a generator; the generated power is dissipated in resistance grids external to the machine or recovered for use. Cradling in trunnion bearings permits the determination of reaction torque. Provision is made to measure torque in either direction, depending on the direction of rotation and mode of operation.

When used as a driving dynamometer, the device performs as a d.c. motor, which presents a problem in certain instances of obtaining an adequate source of d.c. power for this purpose. Use of either an a.c. motor-driven d.c. generator set or a rectified source is required. Ease of control and good performance at slow speeds are features of the d.c. dynamometer. The d.c. dynamometer can be adjusted to provide any torque from zero to the maximum design value for speeds from zero to the so-called base speed of the machine. This is the speed at which the maximum torque develops the maximum design power. At speeds above base speed, torque must be gradually reduced so as to maintain power less than the maximum design value.

(iii) ELECTRIC MOTORS OR GENERATORS. Electric motors or generators may be adopted for use in dynamometry. This is feasible when d.c. rather than a.c. machinery is used. Cradling the motor or generator may be done for either driving or absorbing applications, respectively. By measuring torque reaction and speed, power may be computed. Adjustment of driving speed or absorption load could be provided through control of field current.

Another method of power measurement is the use of speed characteristic of the motor or generator. If a d.c. generator is used as an absorption dynamometer, then

$$\text{Power absorbed by the system} = \frac{\text{output of generator in kilowatts}}{\text{efficiency}}$$

$$= \frac{e \cdot i}{1000(\text{efficiency})}$$

Likewise, if a d.c. motor is used as a driving dynamometer,

$$\text{power input to the system} = \text{power input to motor in kilowatts} \times \text{efficiency}$$

$$= \frac{e \cdot i \times \text{efficiency}}{1000}$$

where e is in volts, and i in amperes.

In both the cases, torque is given by

$$T = \frac{60P}{2\pi N}$$

Both e and i may be measured separately, or a wattmeter may be employed and the electrical power measured directly.

(iv) a.c. MOTORS OR GENERATORS. The a.c. motors or generators may be used for torque or power measurements, but their use is extremely difficult.

4.7 Transmission dynamometers

A convenient method for measuring torque is by means or solid or hollow tubes twisted by applied torque. Under torque loading there will be tensile and compressive strains at the tube surface at 45° to the tube axis as shown in Fig. 4.14 (a). The relationship between strain and torque is given by

$$\epsilon_{45°} = \pm \frac{Tr_0}{\pi G(r_0^4 - r_1^4)}$$

$$\phi = \frac{2Tl}{\pi G(r_0^4 - r_1^4)}$$

where G is the shear modulus, ϕ is the rotation in radians, l is the length of the tube, and r_0 and r_1 are outside and inside radii of the tube.

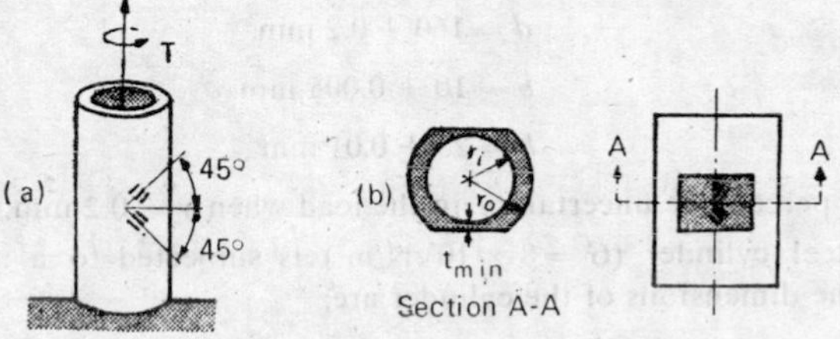

Fig. 4.14 Torque tubes

If the sensitivity obtained from a simple tube is insufficient, it can be increased by thinning the tube wall at the gauging sections as shown in Fig. 4.14 (b). The strain will be raised by the ratio of the initial thickness $(r_0 - r_1)$ to the minimum thickness t_{min}.

In some cases, other elastic elements, such as bar of rectangular cross section, are used. The strain gauges are fixed on the torque sensing element as shown in Fig. 4.15. The gauges are so mounted as to be insensitive to any axial or bending load along with the temperature variations. They measure only the twist. Sometimes optical methods are used to measure deflections. These types of dynamometers are used as a coupling between driving and driven machines, or between any two portions of a machine. In cases where strain gauges are used as secondary transducers, electrical connections are made through slip rings, with means provided to lift the brushes when they are not in use, thus minimising wear. However, slip rings are subject to wear and may

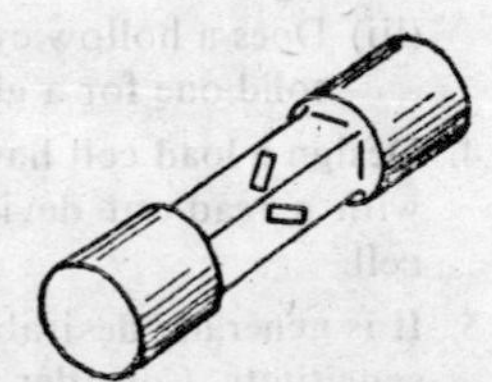

Fig. 4.15 Square shaft torque element

present maintenance problems when permanent installations are required. For this reason many attempts have been made to devise electrical torque meters which do not require direct electrical connections to the moving shaft. Inductive and capacitive transducers are used to accomplish this.

The dynamic response of elastic deflection torque transducers is essentially slightly damped second order, with natural frequency usually determined by the stiffness of the transducer and the inertia of the parts connected at either end.

Exercises

1. A proving ring is constructed of steel ($E = 2 \times 10^6$ N/m²) having a cross section with $b = 10$ mm thickness and $h = 25$ mm depth. The overall diameter of the ring is $d = 160$ mm. A micrometer is used for deflection measurement. It has an uncertainty of ± 2 μm. Assuming that the dimensions of the ring are exact, calculate the applied load when the uncertainty in the load is 1.5%.
   If the dimensions are:
   $$d = 160 \pm 0.2 \text{ mm}$$
   $$b = 10 \pm 0.005 \text{ mm}$$
   $$h = 25 \pm 0.01 \text{ mm}$$
   calculate the percentage uncertainty in the load when $y = 0.2$ mm.
2. A hollow steel cylinder ($G = 8 \times 10^5$ N/m²) is subjected to a moment such that $\phi = 1.50°$. The dimensions of the cylinder are:
   $$r_1 = 12 \pm 0.01 \text{ mm}$$
   $$r_0 = 16 \pm 0.01 \text{ mm}$$
   $$L = 125 \pm 0.03 \text{ mm}$$
   The uncertainty in the angular deflection is $\pm 0.05°$.
   Calculate the nominal value of the impressed moment and its uncertainty. Also calculate the 45° strains.
3. (i) Is a torque tube, with four strain gauges at 45° to the axis, subject to interference from an axial load?
   (ii) Does a hollow torque tube have a higher natural frequency than a solid one for a given strain or torque sensitivity?
   (iii) Does a hollow cylindrical cantilever bar have a higher natural frequency than a solid one for a given strain or load sensitivity?
4. Design a load cell having an electrical output, and a resolution of 0.1 gm when used with a read-out device sensitive to 10^{-5} volt. Estimate the natural frequency of the cell.
5. It is generally desirable, in strain gauge load cell, to have both high stiffness and high sensitivity. Consider three load cells, an axial rod, a cantilever and a strain ring. For a sensitivity of 3 μm/m strain per kg load, using 6 mm × 6 mm gauges, what is the maximum stiffness practically obtainable for the three different types of cells?

5

Measurement of Stress and Strain

5.1 Introduction

When a stress is applied to a body, it gets deformed and these deformations are related to the applied stress. Under the state of loading, it is very desirable to find out the actual stress distribution in the body. The evaluation of stress distribution is known as stress analysis and includes the determination of kinds, magnitudes and directions of the stress. Theoretical stress analysis is possible for simple geometries; however, for complicated geometries, many assumptions are made and the mathematics is too complicated. For stress analysis, stress is not directly measured but obtained indirectly from the measurement of deformations or strains. Therefore an experimental stress analysis is not 100% experimental but involves the use of theoretical relations.

There are a number of methods available for measuring strain. The following methods have been discussed in this chapter:

(i) Mechanical methods
(ii) Opto-mechanical methods
(iii) Electrical strain gauges
  (a) capacitance gauges
  (b) inductive gauges
  (c) piezo electric gauges
  (d) resistance gauges
(iv) Grid method
(v) Moire fringe technique
(vi) Interferometry
(vii) Photo-elasticity
(viii) Holography

5.2 Mechanical methods

Since the magnitude of deformation to be measured is very small, some kind of magnification is to be used. Prior to 1930, all instruments were of mechanical nature and mechanical amplification was achieved with wedges, screws, levers, gears and their combinations. The length of the gauge was as large as 250 mm and hence only a value of deformation averaged over such large distances was obtained. Therefore, these instruments (gauges) were not suited for the steep gradients of strain. Further, it is rather difficult to design a mechanical system which gives a precise and desired magnification. The effects of various factors such as friction, lost motion, the weight and inertia, and flexibility of the parts make it difficult to obtain accurate and reliable results from these instruments. Fig. 5.1 shows one such gauge using lever and dial gauge combination (gears) for amplification. These instruments cannot be used for the study of dynamic strains as are obtained in impact loading, etc.

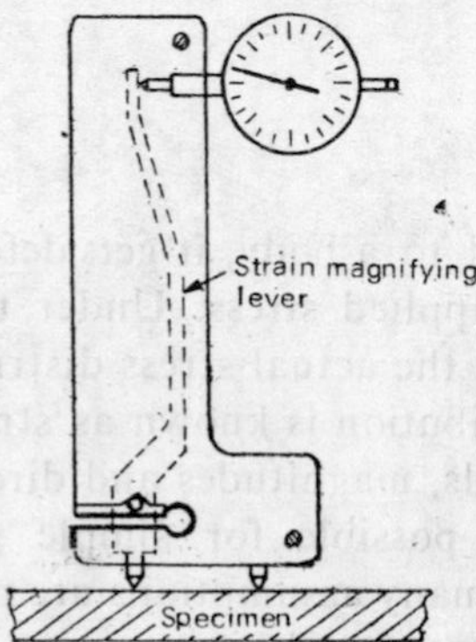

Fig. 5.1 Mechanical strain gauge

Another mechanical gauge where the amplification is achieved by an ingenious way is currently available. This is known as a 'mikrokator' and makes use of a double helix, oppositely twisted. The gauge is very sensitive up to 0.2 μm and the dynamic response is fair.

5.3 Opto-mechanical methods

In these types of gauges, optical levers are used for magnification. Because of high magnification achievable, the optical gauges are much smaller in length about 50 mm. One such gauge is the Tuckerman strain gauge shown in Fig. 5.2. The operation of the Tuckerman gauge is based on the rotation of mirror, which is one of the face of a tungsten-carbide rocker called as the Lozenge by the manufacturer. A beam of collimated light from an auto-collimator is reflected from a mirror prism and a face of Lozenge in the extensometer. The extensometer is placed on the specimen and the mirror prism is so adjusted as to receive the image in the auto-collimator eye-piece at a reference point. When the specimen is deformed, say elongated, the rocker will move outwardly, deflecting the beam and hence the image in the eye-piece to the left. The linear displacement of the image is dependent on the deflection of the rocker and the focal length of the auto-collimator. This relative movement is calibrated as the strain. An important feature of this sort of instrument is that there need not be a fixed relationship between the positions of auto-collimator and the extensometer.

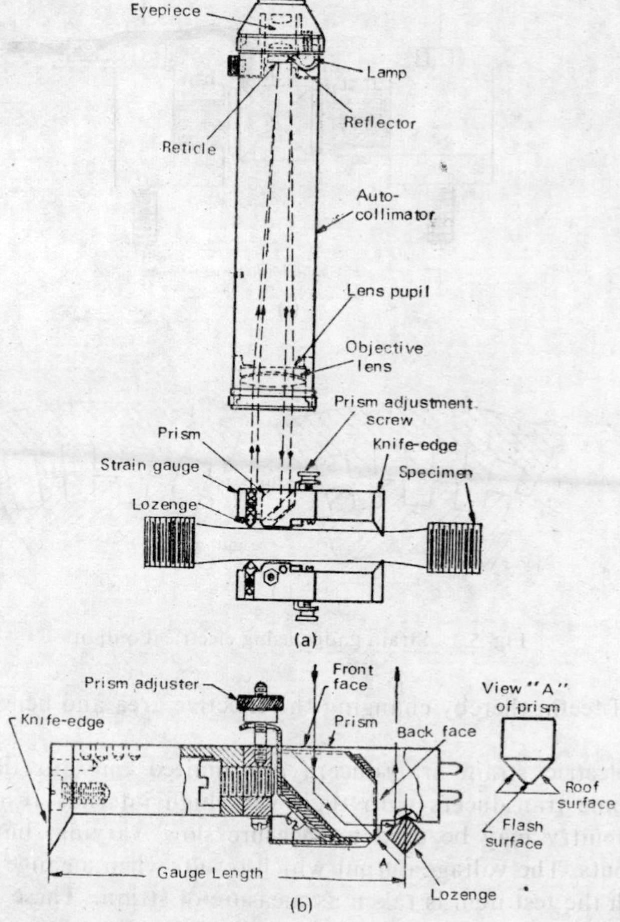

Fig. 5.2 Opto-mechanical strain gauge

5.4 Electrical methods

The electrical methods of measuring strain possess the advantages of dynamical measurement, high sensitivity, etc. The inductive and capacitive gauges are of large mass and size, and are used only for some special applications. These are quite rugged and maintain the calibration over long period of time. These gauges are sometimes used as load indicators, mounted directly on the machine frames. The changes in inductance or capacitance due to strains caused by loading are calibrated in terms of strain. Fig. 5.3 (a) illustrates an inductive gauge for general purpose application. These types of gauges are bolted in place permanently. In one such application of inductive gauge, it is directly mounted on the frame of a rolling-mill for the measurement of rolling loads in steel mill. Fig. 5.3 (b) shows a torque meter using capacitive type strain gauges. Torque carried by an elastic member causes a shift in the relative

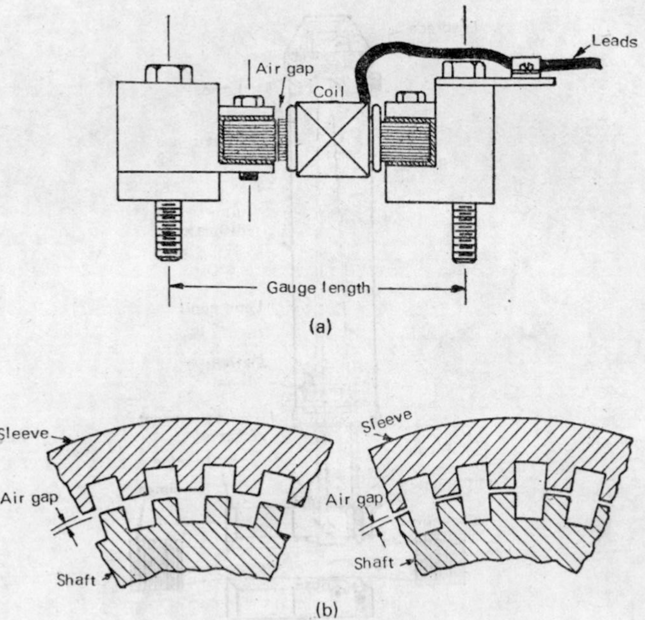

Fig. 5.3 Strain gauge using electrical output

position of teeth, thereby changing the effective area and hence the capacitance.

Piezo electric strain transducers are limited only to the dynamic inputs. Some transducers with their very high internal resistance and proper circuitry may be used to measure slow varying, but rarely for steady inputs. The voltage output which results when a gauge is deformed along with the test item is taken as measure of strain. These gauges are, therefore, to be cemented with the test item. Wafers of barium-titanate 0.25 mm thick with suitable electrodes are bonded to the specimen with DUCO cement. These gauges are equally sensitive to strains in lateral directions and have a very high output sensitivity.

Electrical Resistance-Strain Gauge

Due to their good dynamic response, stability, range of available size, ease of data presentation and processing, etc., resistance type of strain gauges are widely used. Therefore, these gauges are described here.

The principle on which resistance strain gauge works may be called as 'piezo-resistivity'. The resistance of a wire conductor changes when it is strained. The change in the resistance bears a definite relation with the strain or the applied stress. The resistance strain gauges are of two types:

(i) unbonded strain gauges, and
(ii) bonded strain gauges

In unbonded strain gauge, two to twelve loops of high-tensile resistance wire (commercially referred to as alloy 479, containing 92% Pt and 8% W) of about 0.025 mm diameter, are attached to both a stationary frame and a movable platform with the help of insulated pins. Relative motion between the frame and the platform is possible as guided by flexure plates. One such construction is shown in Fig. 5.4. Unbonded gauges are used mainly as elements of force and pressure transducers, and accelerometers rather than for strain measurement.

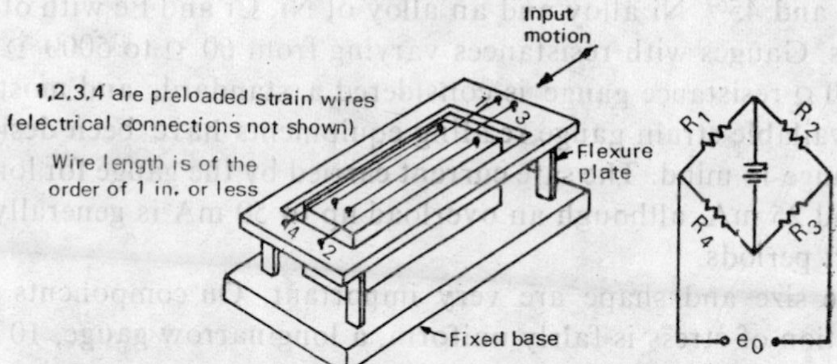

Fig. 5.4 Unbonded strain gauge

On the other hand, bonded type of strain gauges are cemented to the test member; when properly cemented, they effectively form part of the surface to which they are attached and undergo the same strain. A bonded type of strain gauge consists of, as a resistive element, a length of fine metal wire, approximately 0.025 mm in diameter or a metal foil 0.005 mm thick or a whisker of semiconducting material, Fig. 5.5. To reduce the

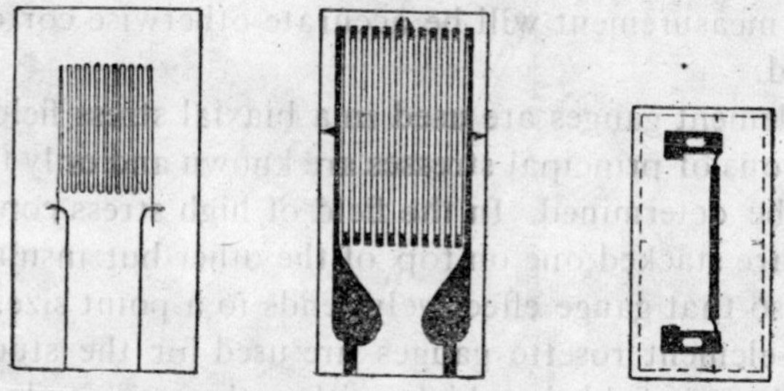

Fig. 5.5 Types of electrical resistance strain gauges

length of the gauge (so that the strain over a very small region is measured rather than the average strain) while retaining its sensitivity (α_R), the wire, or the foil, is usually formed in a grid pattern. The resistive element is fixed to a suitable base from the following:

(i) wire wound around a paper or sandwiched between the paper. Leakage resistance is of the order of 1000 MΩ,
(ii) Bakelite base is used for wire and foil gauges for high temperature applications,
(iii) plastic and rubberised stripable base for foil gauge.

The paper base strain gauge may be cemented with DUCO cement under 0.15 kg/cm^2 pressure and curing takes about 7–8 hours. For Bakelite based strain gauges, phenol resin is employed with a pressure of 1.8 kg/cm^2 during curing. Curing lasts 24 hours. Epoxy resin may be used for plastic base strain gauges. The change of resistance brought about due to the application of load is measured with some form of Wheatstone bridge to be discussed later.

The most common metals used for the manufacture of strain gauges are 55% Cu and 45% Ni alloy and an alloy of Ni, Cr and Fe with other minor elements. Gauges with resistances varying from 60 Ω to 5000 Ω are available; 120 Ω resistance gauge is considered a standard, and most commercially available strain gauge reading equipments have been designed with 120 Ω gauge in mind. The safe current carried by the gauge for long periods is around 25 mA, although an overload up to 50 mA is generally accepted for short periods.

Gauge size and shape are very important. On components where the distribution of stress is fairly uniform, a long narrow gauge, 10 to 25 mm long, can be employed. In regions of likely stress concentration and steep strain gradients, a much shorter gauge length must be used. Gauges are available down to 0.4 mm effective length.

A gauge will only react to strain parallel to the direction of wire. If strain is not parallel to the gauge length, then only that component of strain in the direction of gauge will be recorded. Thus

(i) Single-element gauges are used in uniaxial fields, or for making rosettes. If the gauge is used as prescribed by calibration procedure, strain measurement will be accurate otherwise correction is to be applied.

(ii) Two element gauges are used in a biaxial stress field, in which the directions of principal stresses are known and only their magnitude is to be determined. In the field of high stress concentration, the grids are stacked one on top of the other but insulated from each other so that gauge effectively tends to a point size.

(iii) Three-element rosette gauges are used for the study of a general biaxial stress field, in which neither the magnitude nor the directions of principal stresses are known. The choice of overlapping or a single plane style is determined by the nature of strain gradient at the point of consideration, where the gauge is to be mounted. Fig. 5.6 illustrates single element, two-element stacked over each other and three-element rosette in 60°, and 90°–45° configuration of strain gauges.

Analytical theory of strain gauge: piezo-resistive effect

A simple approach for the theory of strain gauge is discussed here. The resistance R of a wire of length L, area of cross section A and specific

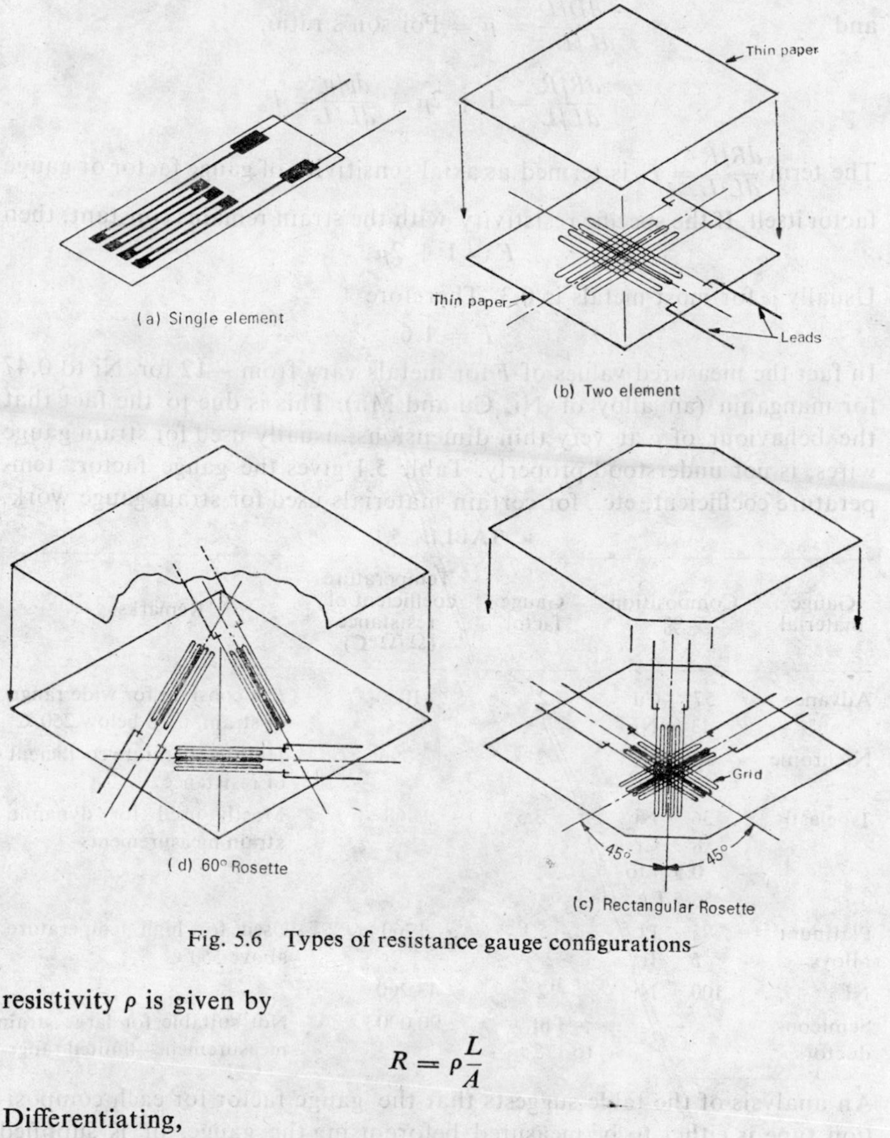

Thin paper

(a) Single element

Thin paper

Leads

(b) Two element

Grid

45° 45°

(d) 60° Rosette

(c) Rectangular Rosette

Fig. 5.6 Types of resistance gauge configurations

resistivity ρ is given by

$$R = \rho \frac{L}{A}$$

Differentiating,

$$\frac{dR}{R} = \frac{d\rho}{\rho} + \frac{dL}{L} - 2\frac{dD}{D}$$

where $A = CD^2$; C is a constant, its value being 1 for square cross section of dimension D and $\pi/4$ for circular cross section of diameter D.

Defining

$$\frac{dL}{L} = \epsilon_a = \text{axial strain},$$

$$\frac{dD}{D} = \epsilon_l = \text{lateral strain},$$

and
$$\frac{-dD/D}{dL/L} = \mu = \text{Poisson's ratio},$$

$$\frac{dR/R}{dL/L} = 1 + 2\mu + \frac{d\rho/\rho}{dL/L} = F_a$$

The term $\frac{dR/R}{dL/L} = F_a$ is termed as axial sensitivity of gauge factor or gauge factor itself. If the specific resistivity with the strain remains constant, then

$$F = 1 + 2\mu$$

Usually μ for most metals is 0.3. Therefore

$$F = 1.6$$

In fact the measured values of F for metals vary from -12 for Ni to 0.47 for manganin (an alloy of Ni, Cu and Mn). This is due to the fact that the behaviour of ρ at very thin dimensions, usually used for strain gauge wires, is not understood properly. Table 5.1 gives the gauge factor, temperature coefficient, etc., for certain materials used for strain gauge work.

TABLE 5.1

Gauge material	Composition %		Gauge factor	Temperature coefficient of resistance (Ω/Ω°C)	Remarks
Advance	57	Cu	2	10.8	F is constant for wide range of strain, used below 250°C
	43	Ni			
Nichrome	80	Ni	2	396	High temperature coefficient of resistance
	20	Cr			
Isoelastic	36	Ni	3.5	468	Mostly used for dynamic strain measurements
	8	Cr			
	0.5	Mo			
	55.5	Fe			
Platinum alloys	95	Pt	5.1	1260	Used for high temperature above 550°C
	5	Ir			
Ni	100	Ni	−12	43,200	
Semiconductors	—		−140 to 175	90,000	Not suitable for large strain measurement—limited range

An analysis of the table suggests that the gauge factor for each composition type is either to be measured before using the gauge, or is supplied by the manufacturer which usually is the case.

If the resistance gauge is strained to the point, that its element is operating in the plastic region, then $\mu = 0.5$ and hence $F = 2.0$. For most commercial strain gauges the gauge factor is the same for both compressive and tensile strains.

Cross Sensitivity

Though most of the gauge element wire is parallel to the axis of the gauge (direction of stress), a certain amount of wire is unavoidably placed transverse to the gauge axis. This length of wire will react to the strains perpendicular to the gauge direction and gives rise to cross-sensitivity, F_l The ratio F_a/F_l is generally about 2%. In foil type gauges, the effect of this length is considerably minimised by decreasing the resistance of the cross element; the foil at the ends is usually of large surface area and hence of low resistance.

The manufacturer's calibration is carried out on a uniaxial stress field, so that if a gauge is used in service in the same manner, cross-sensitivity may be ignored. If a gauge is used in a two dimensional stress field, the readings will be influenced by the transverse strain and a correction should, therefore, be applied to obtain the true strain along the gauge axis. The true strains in the two perpendicular directions ϵ_1 and ϵ_2 are given by

$$\epsilon_1 = \frac{1 - \mu\eta}{1 - \eta^2}(\epsilon_1' - \eta\,\epsilon_2')$$

and

$$\epsilon_2 = \frac{1 - \mu\eta}{1 - \eta^2}(\epsilon_2' - \eta\,\epsilon_1')$$

where ϵ_1' and ϵ are the measured strains using the gauge and $\eta = F_n/F_a$.

Effect of Temperature and Humidity

Variation of room temperature influences the gauge readings in three ways:

(i) a change in the temperature causes a change in the resistance of the gauge element,

(ii) due to the differential expansion between the test member and the gauge bond material, an apparent strain will be induced in the gauge, and

(iii) the gauge factor of a strain gauge gets affected by temperature owing to 'creep'; the variation with time at constant applied stress. Up to a temperature of 200°C, constantan wire gauge with bakelite base and phenol-resin cement has minimum creep.

The above statements can be easily described by the following relationship:

$$\left(\frac{\Delta R}{R}\right)_0 = \epsilon \cdot F_0(1 + m_F\Delta\theta) + \beta\Delta\theta + (\alpha_1 - \alpha_2)F\Delta\theta \qquad (1)$$

where F_0 = gauge factor at room temperature θ_0 and m_F is its temperature coefficient,

β = temperature coefficient of resistance of gauge wire,

α_1, α_2 are the coefficients of linear expansion of test structure and of the gauge wire respectively,

$\Delta\theta = \theta_i - \theta_0$.

If the substrate expands more than the gauge wire at a given temperature, the latter suffers a tensile strain. If $\alpha_2 > \alpha_1$, the gauge wire will suffer compression.

The methods of achieving partial or complete temperature compensation over a limited range of measured strain values are:

(i) by making

$$\beta + K(\alpha_1 - \alpha_2) = 0$$

For metals β is positive and hence this equation can be satisfied by choosing $\alpha_2 > \alpha_1$ suitably.

(ii) by using series combination of two 'opposing' wire materials, such as constantan and nickel (dual gauges), the latter having a negative gauge factor. This method aims at making the entire sum in equation (1) equal to F_ϵ over a limited range of strain values.

(iii) by locating a temperature-sensitive resistor close to gauge and connecting it electrically in series with the bridge supply line so that the sensitivity (output voltage/input strain) remains unaltered.

The most common and the practical way of compensating temperature errors is either by the use of dummy gauges or by employing wire resistance gauges in push-pull pairs. Examples of the various arrangements for compensating temperature errors will be given at the later section of this chapter.

Humidity is another factor which may seriously affect the performance of strain gauges. Corrosion of the gauge wires results in an increase in its resistance. The effect of humidity can be minimised by giving a coat of wax on the bonded strain gauge.

Strain Gauge Sensitivity Equation

Figure 5.7 shows a Wheatstone bridge circuit with four active gauges of resistances R_1, R_2, R_3, and R_4, and excited by the voltage V. Initially the bridge is balanced, i.e.

$$\frac{R_1}{R_2} = \frac{R_4}{R_3}$$

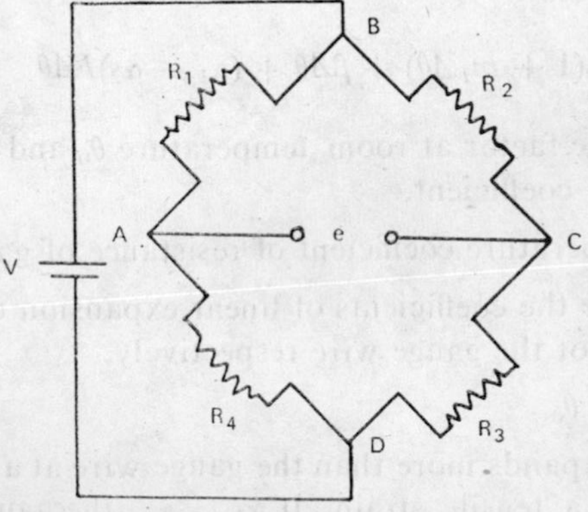

Fig. 5.7 Wheatstone bridge circuit

When the stress is applied, the resistance of gauges will change by ΔR_i ($i = 1, \ldots, 4$) and hence a potential difference e will be developed across AC. The potential difference e will be

$$e = V\left[\frac{R_2\, dR_1}{(R_1 + R_2)^2} - \frac{R_1\, dR_2}{(R_1 + R_2)^2} + \frac{R_4\, dR_3}{(R_3 + R_4)^2} - \frac{R_3\, dR_4}{(R_3 + R_4)^2}\right]$$

Writing this equation in terms of gauge factor F and strains ϵ (and assuming $R_1 = R_2 = R_3 = R_4 = R$) the output voltage will be given by

$$e = \frac{VF}{4}(\epsilon_1 - \epsilon_2 + \epsilon_3 - \epsilon_4) = \frac{VF}{4}\epsilon_{\text{net}}$$

This is known as sensitivity equation. Based on this equation and requirement of temperature compensation, some arrangements for mounting strain gauges for some specific purposes will now be described.

1. *Positioning of gauges to measure axial strain*

Figures 5.8 (a) and (b) show a bar mounted with strain gauges and the schematic of bridge circuitry. The gauges R_1 and R_3 are mounted axially on the bar subjected to either tensile or compressive stress; gauges R_2 and R_4 are unstrained. Therefore,

$$\epsilon_2 = \epsilon_4 = 0$$

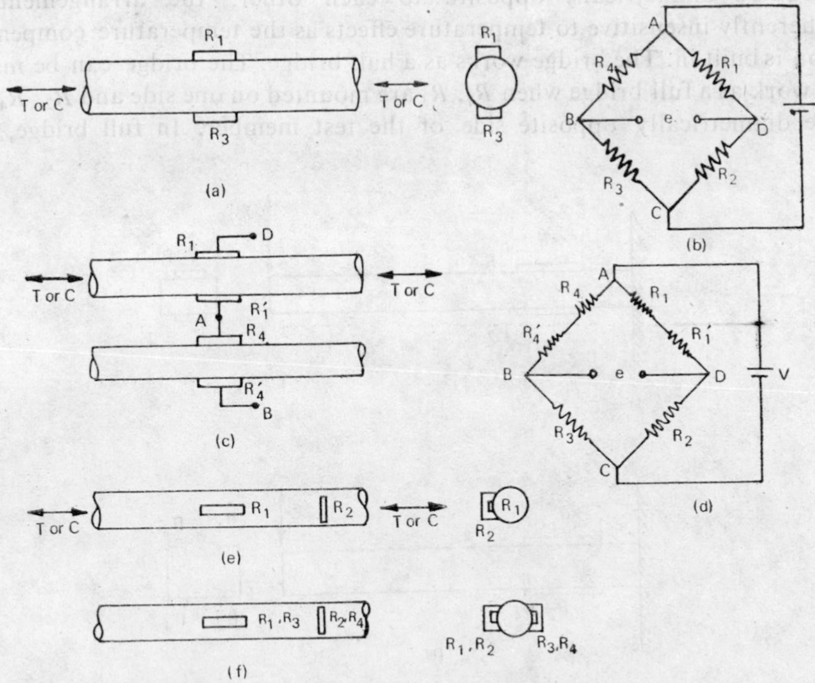

Fig. 5.8 Positioning of strain gauge to measure axial strain

and
$$\epsilon_1 = \epsilon_3 = \epsilon$$

Thus $e = \frac{VF}{4} 2\epsilon$. The arrangement provides double the sensitivity. Further the arrangement is insensitive to bending stresses, as one of the gauges is compressed ($-\Delta R$), while the other is elongated (ΔR), thus cancelling the effects provided they are bonded diametrically opposite to each other.

In another arrangement, where temperature compensation as well as the removal of influence of bending strains are achieved is shown in Figs. 5.8 (c) and (d). The resistances R_4, R_4' are identical to R_1, R_1' and are used only for temperature compensation as dummy gauges. R_1, R_1' are bonded diametrically opposite to each other for annulling the effects of bending strains.

Another very simple arrangement, but mostly recommended for foil gauges ($\simeq$ zero cross-sensitivity) is shown in Figs. 5.8 (e) and (f). In this arrangement, temperature compensating gauges are mounted perpendicular to the axis and hence are not subjected to axial strains. Bending strains are avoided by mounting R_1, R_3 and R_2, R_4 diametrically opposite to each other. The sensitivity is double and the bridge is said to be working as half bridge.

2. *Positioning of gauges to measure bending strains*

Figures 5.9 (a) and (b) show the arrangement. The gauges R_1, R_2 are mounted diametrically opposite to each other; the arrangement is inherently insensitive to temperature effects as the temperature compensation is built in. The bridge works as a half bridge. The bridge can be made to work as a full bridge when R_1, R_3 are mounted on one side and R_2, R_4 on the diametrically opposite side of the test member. In full bridge, the

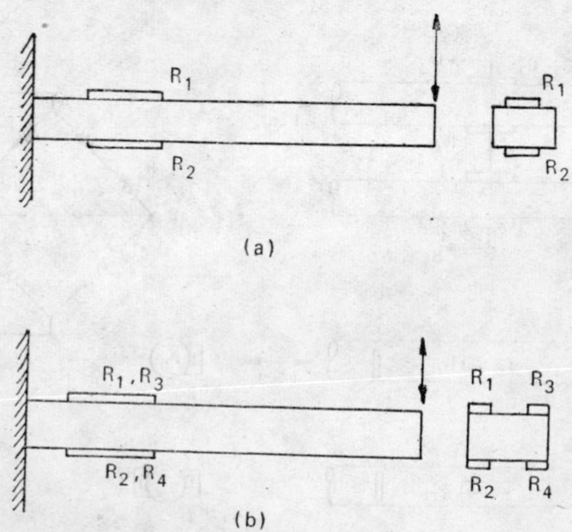

Fig. 5.9 Mounting of gauges to study bending

sensitivity is four times than achievable with a single gauge.

3. *Positioning of gauges to measure torsional strains*

A cylindrical bar subjected to torsion has principal strains at 45° to the longitudinal axis of the bar. Bending and axial strains can be eliminated by the use of gauges as shown in Fig. 5.10. Since the principal axes are

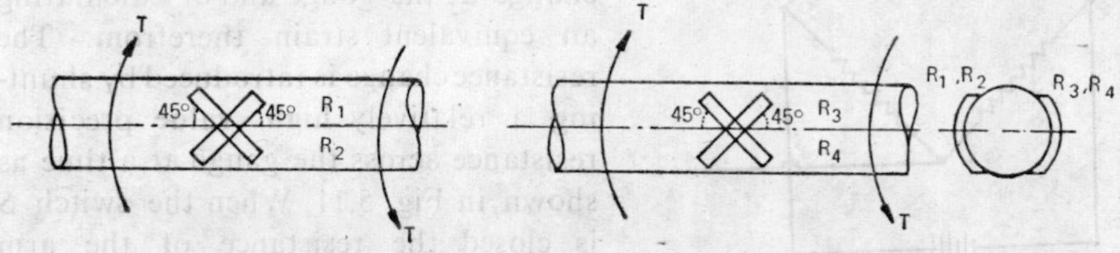

Fig. 5.10 Positioning of gauges to measure torsion

at 45° from the longitudinal axis of the cylinder due to torsion as shown, the resistance of R_2 will increase $(+\Delta R_2)$ and of R_1 decrease $(-\Delta R_1)$, thus pulling the effect. However, resistance change due to axial strain and temperature will be positive $(+\Delta R_1, +\Delta R_2)$ and thus compensate each other, being at the same side of point of null potential. Same holds good for R_3 and R_4. Further R_1, R_2 and R_3, R_4 are diametrically placed, thus the effect of bending will be compensated.

4. *Mounting of gauges in pressure pickups*

One very favourable situation occurs in diaphragm type pressure pickup where both compressive and tensile stresses coexist. The gauges are arranged such that the bridge acts as a full bridge and temperature effects are compensated by push-pull action of gauges.

Calibration

Strain gauges may be used under both the static and dynamic conditions. Hence both static and dynamic calibration procedures shall now be discussed.

STATIC CALIBRATION. Ideally, calibration of any measuring system consists of introducing an accurately known sample of the variable that is to be measured and then observing the system's response. This ideal cannot often be realised in bonded resistance gauge work because of the nature of transducer. Normally the gauge is bonded to a test item for the simple reason that the strains are unknown. Once bonded, the gauge can hardly be transferred to known strain situation for calibration. Thus when the gauge is used for the purpose of experimentally determining strains, however, some other approach to the calibration problem is required.

Resistance strain gauges are manufactured under carefully controlled conditions, and the gauge factor for each lot of gauges is provided by the manufacturer with an indicated tolerance of about $\pm 0.2\%$. Knowledge of gauge factor and gauge resistance makes possible a simple method of calibrating any resistance strain gauge system. The method consists of determining the system's response to the introduction of a known small resistance change at the gauge and of calculating an equivalent strain therefrom. The resistance change is introduced by shunting a relatively high value precision resistance across the gauge at a time as shown in Fig. 5.11. When the switch S is closed the resistance of the arm containing R_1 is changed by a small amount ΔR, given by

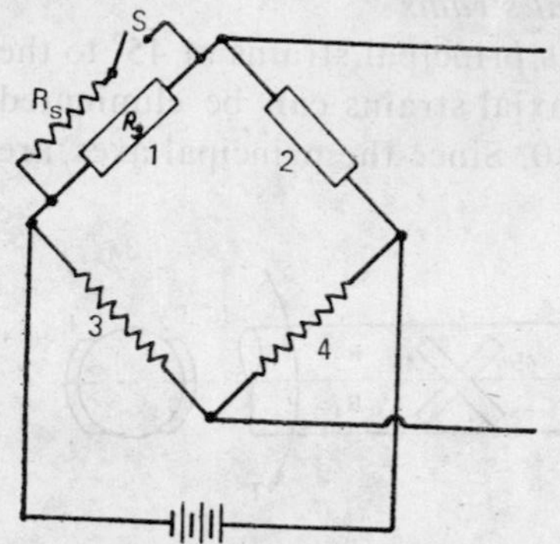

Fig. 5.11 Wheatstone bridge circuit

$$\Delta R = \frac{R_1^2}{R_1 + R_{c1}}$$

where R_{c1} is the calibration resistance. Therefore the equivalent strain is

$$\epsilon_{c1} = \frac{1}{F} \cdot \frac{\Delta R}{R_1} = \frac{1}{F} \frac{R_1}{R_1 + R_{c1}}$$

A graph between resistance change ΔR and equivalent strain can be plotted as strain read off from the measured resistance changes when the gauge is bonded on the test member.

DYNAMIC CALIBRATION. Dynamic calibration is sometimes provided by replacing the manual calibration switch with an electrically driven switch, often referred to as *chopper*, which makes and breaks the contact 60 to 100 times per second. When displayed on CRO screen or recorded, the trace obtained is found to be a square wave. The step in the trace represents the equivalent strain calculated from

$$\epsilon_c = \frac{1}{F} \cdot \frac{R_1}{R_1 + R_{c1}}$$

5.5 Measurement of strains

The bridge can be excited either by a direct current or an alternating current. Further the strain to be measured could be of static, slowly varying or dynamic nature. However, most of the metal forming processes are subjected to dynamic strain. For static strains, a galvanometer can be used for measurements. However, its use for dynamic strains is restricted due to the high inertia of moving parts and hence it is substituted by pen recorder or CRO which possesses the necessary dynamic range. Further, the output of the bridge is very small, and hence a high gain d.c. amplifier for static or slowly varying strains is required as shown in Fig. 5.12. Building of a high gain amplifier to amplify such low level

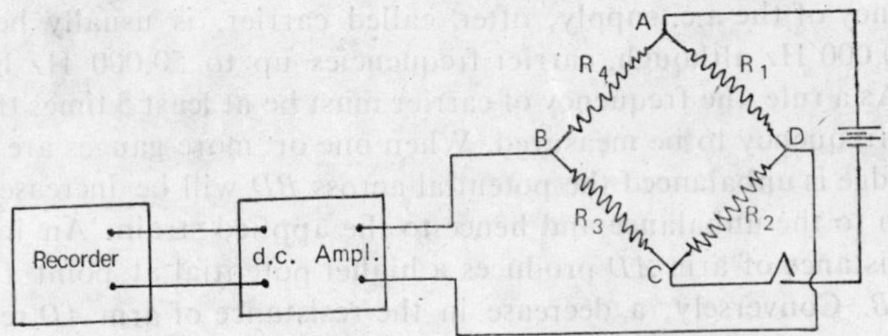

Fig. 5.12 Recording of slow varying or static strains

signals of very low frequency has the following problems associated with it:

(i) small changes in the terminal potentials at different stages due to d.c. drift will be of the same order as of the unbalance voltage and will be amplified along with the signal,

(ii) contact potentials and thermal emf's in the various parts of the circuit may be comparable to the unbalance voltage,

(iii) stray potential of 50 Hz from the vicinity of the long cables used for measurements.

On the other hand a.c. amplifier, though suitable for dynamic measurement cannot be used for static or slowly varying signals. A method using a carrier frequency for both slow varying and dynamic strains will now be described.

a.c. Excited Bridge

In this arrangement, the bridge is excited by alternating current as shown in Fig. 5.13. The a.c. is applied across the terminals A and C; the

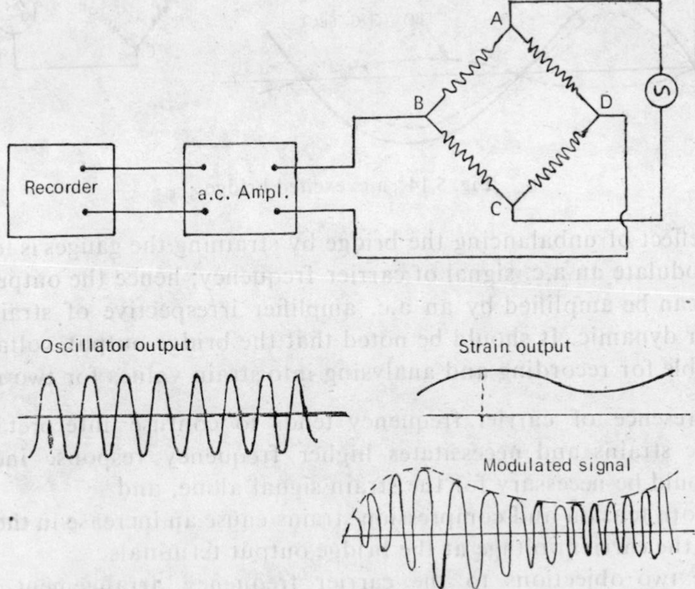

Fig. 5.13 a.c. excitation of bridge

frequency of the a.c. supply, often called carrier, is usually between 50 and 10,000 Hz although, carrier frequencies up to 50,000 Hz have been used. As a rule, the frequency of carrier must be at least 5 times the highest strain frequency to be measured. When one or more gauges are strained, the bridge is unbalanced the potential across *BD* will be increased in proportion to the unbalance and hence to the applied strain. An increase in the resistance of arm *AD* produces a higher potential at point *D* than at point *B*. Conversely, a decrease in the resistance of arm *AD* results in a higher potential at *D* than at *B*. Therefore, tension and compression strains produce voltages of opposite sign at the output terminals of the bridge. Since the bridge is excited by alternating voltage, both tension and compression increase the alternating voltage at the output. However, if this alternating voltage is considered as the difference between the voltage across *BC* and the voltage across *DC*, an increase in the resistance of arm *AD* produces a resultant bridge voltage which is in phase with both the voltage across *DC* and that across *BC*, whereas the decrease in the resistance of arm *AD* produces a resultant bridge voltage out of phase with both voltage *DC* and voltage *BC* as shown in Fig. 5.14. Therefore, tension and compression produce alternating voltages of opposite phases at the bridge output.

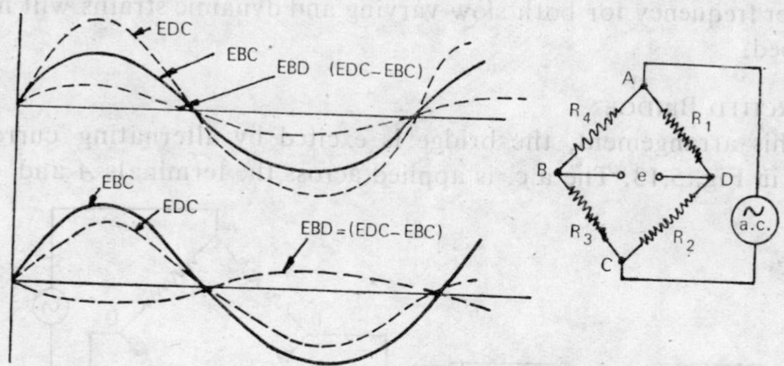

Fig. 5.14 a.c. excited bridge

The effect of unbalancing the bridge by straining the gauges is to amplitude modulate an a.c. signal of carrier frequency; hence the output of the bridge can be amplified by an a.c. amplifier irrespective of strain being static or dynamic. It should be noted that the bridge output voltages are unsuitable for recording and analysing into strain values for two reasons:

(i) presence of carrier frequency tends to confuse interpretation of dynamic strains and necessitates higher frequency response indicators than would be necessary for the strain signal alone, and

(ii) both tension and compression strains cause an increase in the amplitude of the carrier voltage at the bridge output terminals.

These two objections to the carrier frequency arrangement can be eliminated in either of the two ways:

(i) The first method shown in Fig. 5.15 consists of rectifying and filtering the amplitude modulated carrier signal after amplification, thus eliminating the carrier voltage. Tensile and compressive strains can be detected by purposefully unbalancing the bridge to such a point that the maximum anticipated strain will not bring the bridge back to the balanced point. It should be remembered that the sensitivity adjustment of the amplifier change the zero point of the indicating instrument.

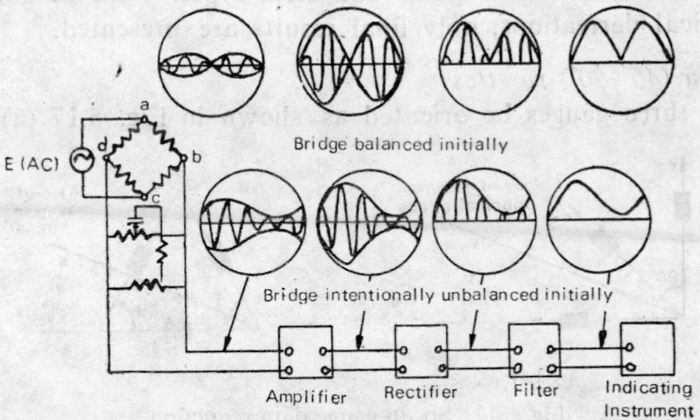

Fig. 5.15 An a.c. excited bridge; method of intentional unbalancing

(ii) This method, usually preferred in the laboratory, utilizes a demodulator so that the tensile and compressive strains can be detected without initially unbalancing the bridge. Fig. 5.16 illustrates the schematic arrangement. The demodulator has the advantage that its output is of one polarity when the amplifier output is in phase with the original carrier voltage and is of opposite polarity when the amplifier output is out of phase with the carrier voltage. Thus, effective phase change caused by going from tensile to compressive strains on the bridge is easily detected. For measurements of varying strain, the final signal is used to drive some form of deflection type indicator. In the case of static or slowly varying strains, the signal may be displayed on a deflection type indicator for direct

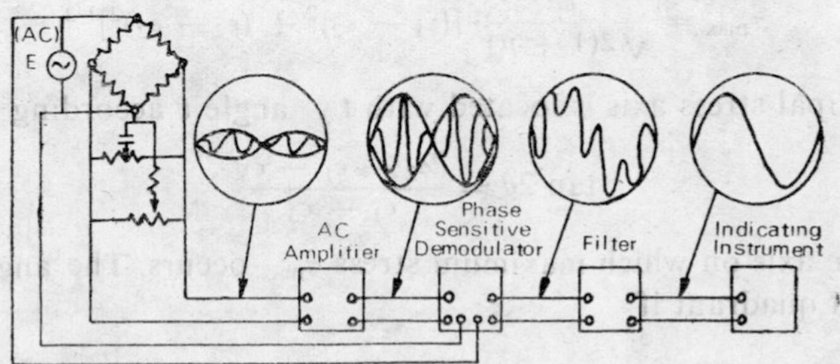

Fig. 5.16 An a.c. excited bridge; use of phase sensitive demodulator

reading or as a simple indication of unbalance for use in a null system.

5.6 Strain gauge data reduction

The aim of the strain measurement is to obtain magnitude and direction of principal strains and hence of principal stresses. An arrangement of strain gauges, often called 'rosettes', is used for this purpose. Two most common rosettes are rectangular and delta types. Without going to the mathematical derivations, only final results are presented.

Rectangular (45°–90°) rosettes

Let the three gauges be oriented as shown in Fig. 5.17 (a) and the

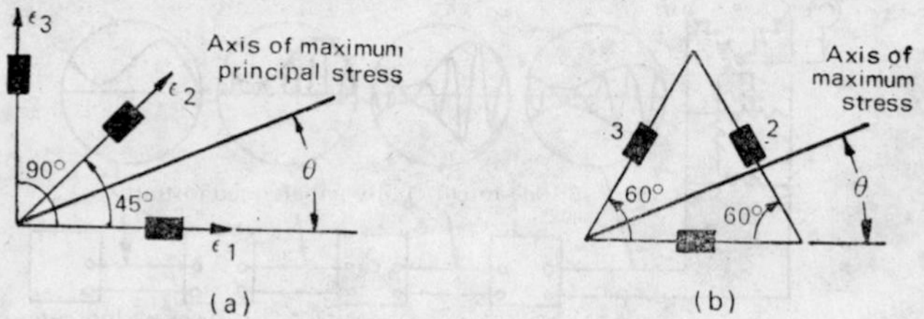

Fig. 5.17 Strain gauge data reduction

measured strains from these gauges be ϵ_1, ϵ_2 and ϵ_3. The principal strains for this situation are

$$\epsilon_{max} = \frac{\epsilon_1 + \epsilon_2}{2} + \frac{1}{\sqrt{2}} [(\epsilon_1 - \epsilon_2)^2 + (\epsilon_2 - \epsilon_3)^2]^{1/2}$$

and
$$\epsilon_{min} = \frac{\epsilon_1 + \epsilon_2}{2} - \frac{1}{\sqrt{2}} [(\epsilon_1 - \epsilon_2)^2 + (\epsilon_2 - \epsilon_3)^2]^{1/2}$$

The principal stresses are

$$\sigma_{max}, \sigma_{min} = \frac{E(\epsilon_1 + \epsilon_2)}{2(1 - \mu)} \pm \frac{E}{\sqrt{2}(1 + \mu)} [(\epsilon_1 - \epsilon_2)^2 + (\epsilon_2 - \epsilon_3)^2]^{1/2}$$

The maximum shear stress τ_{max} is calculated from

$$\tau_{max} = \frac{E}{\sqrt{2}(1 + \mu)} \cdot [(\epsilon_1 - \epsilon_2)^2 + (\epsilon_2 - \epsilon_3)^2]^{1/2}$$

The principal stress axis is located with t   angle θ according to

$$\tan 2\theta = \frac{2\epsilon_2 - \epsilon_1 - \epsilon_3}{\epsilon_1 - \epsilon_3}$$

This is the axis on which maximum stress σ_{max} occurs. The angle θ will be in the first quadrant if

$$\epsilon_2 < \frac{\epsilon_1 + \epsilon_3}{2}$$

and in the second quadrant, if

$$\epsilon_2 > \frac{\epsilon_1 + \epsilon_3}{2}$$

Delta (60°–60°) rosettes

The arrangement of gauges in this case is shown in Fig. 5.17 (b). The principal strains in this situation are given by

$$\epsilon_{max}, \epsilon_{min} = \frac{\epsilon_1 + \epsilon_2 + \epsilon_3}{3} \pm \frac{\sqrt{2}}{3}[(\epsilon_1 - \epsilon_2)^2 + (\epsilon_2 - \epsilon_3)^2 + (\epsilon_3 - \epsilon_1)^2]^{1/2}$$

The principal stresses are

$$\sigma_{max}, \sigma_{min} = \frac{E(\epsilon_1 + \epsilon_2 + \epsilon_3)}{3(1 - \mu)} \pm \frac{\sqrt{2}E}{3(1 + \mu)}[(\epsilon_1 - \epsilon_2)^2 + (\epsilon_2 - \epsilon_3)^2 + (\epsilon_3 - \epsilon_1)^2]^{1/2}$$

The maximum shear stress is calculated from

$$\tau_{max} = \frac{\sqrt{2}\,E}{3(1 + \mu)}[(\epsilon_1 - \epsilon_2)^2 + (\epsilon_2 - \epsilon_3)^2 + (\epsilon_3 - \epsilon_1)^2]^{1/2}$$

The principal stress axis is located according to

$$\tan 2\theta = \frac{\sqrt{3}(\epsilon_3 - \epsilon_2)}{2\epsilon_1 - \epsilon_2 - \epsilon_3}$$

The angle θ will be in the first quadrant when

$$\epsilon_3 > \epsilon_2$$

and in the second quadrant when

$$\epsilon_3 < \epsilon_2$$

5.7 Grid technique

The method consists in placing reference marks on the specimen over the area of interest, measuring the distance between the marks before and after straining and computing the strain as the change in length divided by the original length. In this arrangement, the distance between the marks is the gauge length over which the strain is averaged. The reference marks are in the form of a continuous grid pattern; rectangular, polar or of other forms. Sufficient information is made available from these patterns to determine magnitudes and directions of principal stress at every point. For example, with a rectangular grid, the normal strain in these directions can be determined by the measurement of the change in length of sides and a diagonal. With the normal strain in three directions known, the principal strains can be determined by using the usual rosette equation.

The grid patterns may be formed on the surface of the test member by several methods, viz.

(i) drawing or scribbling,
(ii) photographic printing,
(iii) photo-etching, or
(iv) by cementing a grid network (photographically made) on the surface.

The coarse grids can be made by drawing or scribbling and are used only for the measurement of large strains—in cases of large elastic deformations. For good accuracy, the finer grid patterns are used, with finer grids photographically made. Photo grids have been found to possess remarkable tenacity and have been applied to metal specimen which were later deep drawn.

The necessary grid dimensions can be measured in a number of ways. In some cases, the measurements can be made on the model using a micrometer eyepiece or a microscope with a reticle. Care must be taken to ensure that magnification is the same when readings are taken before and after straining. More often, photographs with high resolution and distortion-free lens are made as the test progresses and evaluated leisurely. The effect of film shrinkage due to wet photographic process should be taken into account.

When the measurements are to be made on a highly curved surface, a corner, etc. Miller's replica technique can be used. This technique is very laborious, requiring measurements at every grid square. The influence of dimensional changes can be made visible by the technique of Moiré fringes.

5.8 Moiré fringe technique

The Moiré fringe effect is the formation of alternately bright and dark bands when two patterns (usually line gratings) are laid on top of the other and one rotated relative to other. For strain measurement, one of the gratings is mounted/projected on the test member, which undergoes the deformation identical to the test member. This grating is called a test grating. The deformed grating is compared with the second grating, called the master grating. The comparison is carried out over the whole field and hence Moiré technique is called a whole field method. One can use Moiré technique both for in-plane and out-of-plane deformation measurements.

The gratings used for Moiré work are made by equal dark and transparent regions; the width of the dark and transparent regions need not be equal but this gives maximum sensitivity. The Moiré gratings often used have a frequency of 40 lines/mm; sometimes up to 80 lines/mm gratings are also used. The frequency higher than this, when used, should incorporate diffraction phenomenon when analysing the formation of Moiré fringes.

IN-PLANE DEFORMATION

The application of Moiré technique to the in-plane deformation measurements is dealt under the following three categories:

(i) Rigid body motion,
(ii) Deformation in one direction,
(iii) Rotation.

(i) *Rigid body motion*

Consider two Moiré gratings of equal pitch (the frequency of the gratings is same; inverse of frequency is pitch) placed one over the other. As one grating moves over the other, starting from a position when the two gratings are laid in such a way that the dark element of one falls over the bright element of the other so that there is no transmission, the transmission will increase over the whole field. If this two-grating arrangement is illuminated by a collimated light, and transmitted light is collected and fed to a photo sensor, the output of the photo sensor will be periodic with a period equal to the period of the grating. Therefore, by counting the periods of the output from a fixed datum the rigid body motion can be obtained easily. Note, initially the gratings are set parallel to each other and they remain so during the measurement.

(ii) *Measurement of deformation along one axis*

The grating on the test member is mounted in such a manner that the grating elements are perpendicular to the direction of deformation. Therefore, loading of the test member will result in the change of period of the test grating, which is compared with the master grating initially set parallel to the former. The experimental arrangement is shown in Fig. 5.18. Let the period of master grating be p and that of deformed grating be p'. Because of this mismatch, Moiré fringes will be formed; the distance d between two Moiré fringes will be governed by the deformation such that the total deformation from a reference which gives the position of a fringe, to the next fringe is one period of the grating. Let there be m grating periods of master grating that fall between two Moiré fringes, then

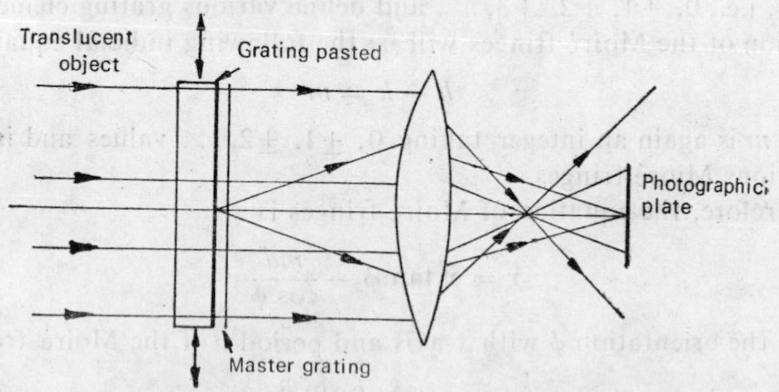

Fig. 5.18 An arrangement to observe Moiré fringes

$$d = mp = (m \pm 1)p'$$

or
$$p' = \frac{m}{m \pm 1} p = \frac{d}{\frac{d}{p} \pm 1}$$

writing $p' = p(1 \pm \epsilon)$, where ϵ is the strain,

$$\epsilon = + \frac{p}{d - p} \text{ for tensile strain,}$$

and
$$\epsilon = - \frac{p}{d + p} \text{ for compressive strain.}$$

The fringe direction in this case is parallel to the grating element and both compressive and tensile strains are calculated from the knowledge of p and the measurement of d. Usually $p \ll d$, and hence equal compressive and tensile strains produce identical fringe structure, and

$$\epsilon = \pm \frac{p}{d}$$

In order to determine the total displacement at any point, the number of Moiré fringes from a known datum is to be found out.

(iii) *Pure rotation of two Moiré gratings*

In this case consider two gratings of different frequencies; this situation is identical to the one where one of the two identical gratings has undergone a linear deformation. Later on, the general result will be discussed for identical gratings.

Let one of the gratings be with period b and elements parallel to x-axis, while the other with period a and elements inclined at an angle θ with the x-axis. The equations of these gratings are:

$$y = bh$$

and
$$y = x \tan \theta + \frac{ak}{\cos \theta}$$

where the integer constants h and k can take both negative and positive values, i.e., 0, ± 1, ± 2, ± 3, . . . and define various grating elements. The equation of the Moiré fringes will be the following indicial equation:

$$h - k = m$$

where m is again an integer taking 0, ± 1, ± 2, . . . values and is a label to various Mioré fringes.

Therefore, the equation of Moiré fringes is

$$y = x \tan \phi - \frac{md}{\cos \phi}$$

where the orientation ϕ with x-axis and period d of the Moiré fringes are

$$\tan \phi = \frac{b \sin \theta}{-a + b \cos \theta}$$

and
$$d = \frac{ab}{(a^2 + b^2 - 2ab \cos \theta)^{1/2}}$$

For a special case, when one of the two identical gratings has suffered a pure rotation of θ,

$$\phi = \theta/2$$

and
$$d = \frac{a}{2 \sin \theta/2} = \frac{p}{2 \sin \theta/2}$$

Thus the Moiré fringes are bisection of the two grating elements. For small angle θ,

$$\theta = p/d$$

Since shear strain γ_{xy} results in rotation, Moiré fringes can be used for the shear strain measurement. If the rotation is α, then the shear strain $\gamma_{xy} = \tan^{-1} \alpha$.

If the rotation due to shear along the x-direction is $(\alpha/2)_x$, then it can be measured by having both gratings parallel to x-axis; then

$$\theta_x = (\alpha/2)_x = 2 \sin^{-1} \left|\frac{p}{2d_x}\right|$$

Similarly,
$$\theta_y = (\alpha/2)_y = 2 \sin^{-1} \left|\frac{p}{2d_y}\right|$$

and
$$\gamma_{xy} = \tan^{-1} [(\alpha/2)_x + (\alpha/2)_y]$$

The relationship between strain and displacement can now be established when the test grating is assumed to undergo both displacement and rotation. Assuming a cartesian reference (Fig. 5.19) system of axes x and y, the component of displacement parallel to x-axis is represented by u and the component parallel to y-axis by v. Moiré fringes representing

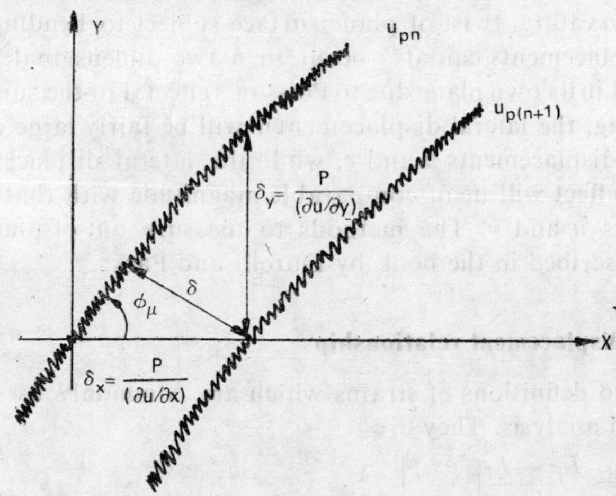

Fig. 5.19 Gradient components of the isothetic U-family along coordinate axes

either u or v family of displacements are called isothetics. Isothetics are, therefore, curves of locus of points that have the same value of displacement component. The displacement $U_i(x, y)$ in a two-dimensional continuum medium may be represented by a surface which can be defined as $Z = U_i(x, y)$. The gradient of the surface $Z = U_i$ will be the slope of the surface in the direction normal to the Moiré fringes. Since the height between two Moiré fringes in the z-direction is p and the period of Moiré fringes is d in the x-y plane, then the gradient

$$\frac{dU_i}{dn} = \frac{p}{d}$$

Let dx and dy be the distances between Moiré fringes in x- and y-directions. The change in displacement component u for points belonging to two adjacent fringes is p. Therefore,

$$\Delta u = p$$

$$\Delta x = dx$$

$$\Delta y = dy$$

In the limit

$$\frac{\partial u}{\partial x} = \frac{p}{dx}$$

and

$$\frac{\partial v}{\partial y} = \frac{p}{dy}$$

Therefore, the strain in a general way, may be defined in terms of the derivatives of the displacement at a point.

OUT-OF-PLANE MEASUREMENT

Moiré method can also be used for the experimental determination of deflection, curvature, twist of plane surface subject to bending, etc. Out-of-plane displacements can also occur in a two-dimensional plane stress model loaded in its own plane due to Poisson's effect. For the surface subjected to bending, the lateral displacement w will be fairly large compared to the in-plane displacements u and v, while the lateral displacement w due to Poisson's effect will be of comparable magnitude with that of in-plane displacements u and v. The methods to measure out-of-plane displacements are described in the book by Durelli and Parks.

5.9 Strain-displacement relationship

There are two definitions of strains which are commonly used in experimental strain analysis. They are

$$\left.\begin{aligned} \epsilon^L &= \frac{L_f - L_i}{L_i}\bigg|_{L_i \to 0} \\ \gamma^L &= (\pi/2 - \xi_f) \end{aligned}\right] \text{ Lagrangian description of strain}$$

and $$\left.\begin{aligned}\epsilon^E &= \frac{L_f - L_i}{L_f}\bigg|_{L_f \to 0} \\ \gamma^E &= (\xi_i - \pi/2)\end{aligned}\right]\ \text{Eulerian description of strain}$$

where ϵ and γ are the direct and shear strains,
L_f and L_i are the final and initial lengths,
ξ_f final angle which was initially a right angle, and
ξ_i initial angle which after deformation becomes a right angle.

The above equations indicate that the difference between the two definitions lies in the choice of base length or base angle. General strain-displacement relations under both descriptions of strain without any mathematical derivation are given here. A detailed analysis may be found in the book by Parks and Durelli. These relations are:

$$\epsilon_x^L = \sqrt{\left[1 + 2\frac{\partial u}{\partial x} + \left(\frac{\partial u}{\partial x}\right)^2 + \left(\frac{\partial v}{\partial x}\right)^2\right]} - 1$$

$$\epsilon_y^L = \sqrt{\left[1 + 2\frac{\partial v}{\partial y} + \left(\frac{\partial v}{\partial y}\right)^2 + \left(\frac{\partial u}{\partial y}\right)^2\right]} - 1$$

$$\gamma_{xy}^L = \sin^{-1}\frac{\dfrac{\partial u}{\partial y} + \dfrac{\partial v}{\partial x} + \left(\dfrac{\partial u}{\partial x}\right)\left(\dfrac{\partial u}{\partial y}\right) + \left(\dfrac{\partial v}{\partial y}\right)\left(\dfrac{\partial v}{\partial x}\right)}{(1 + \epsilon_x^L)(1 + \epsilon_y^L)}$$

and $$\epsilon_x^E = 1 - \sqrt{\left[1 - 2\frac{\partial u}{\partial x} + \left(\frac{\partial u}{\partial x}\right)^2 + \left(\frac{\partial v}{\partial x}\right)^2\right]}$$

$$\epsilon_y^E = 1 - \sqrt{\left[1 - 2\frac{\partial v}{\partial y} + \left(\frac{\partial v}{\partial y}\right)^2 + \left(\frac{\partial u}{\partial y}\right)^2\right]}$$

$$\gamma_{xy}^L = \sin^{-1}\frac{\dfrac{\partial u}{\partial y} + \dfrac{\partial v}{\partial x} - \left(\dfrac{\partial u}{\partial x}\right)\left(\dfrac{\partial u}{\partial y}\right) - \left(\dfrac{\partial v}{\partial x}\right)\left(\dfrac{\partial v}{\partial y}\right)}{(1 - \epsilon_x^E)(1 - \epsilon_y^E)}$$

The terms like $\dfrac{\partial u}{\partial y}, \dfrac{\partial v}{\partial x}$ are known as cross derivatives. The relations under particular cases for example small rotations $\left(\left(\dfrac{\partial u}{\partial y}\right)^2 \to 0, \left(\dfrac{\partial v}{\partial x}\right)^2 \to 0\right)$, small strains, etc., may be obtained easily from the general relations.

There are a number of methods to obtain strains from the Moiré pattern but all of them are tedious and time-consuming. However, a qualitative picture of the whole test field is obtained in a single observation. The sensitivity of the Moiré method depends on the grating frequency and the usual gratings have a frequency of 40 to 80 lines per mm. Higher sensitivity can be obtained either by fringe shifting or mismatch technique. These gratings are obtainable either as stripable photo grating, or may be photo-etched on the specimen. The main advantage of Moiré method is quick appraisal of information and wide range of deformations which could be measured.

5.10 Interferometry

This is also a whole field method and is not often used because of its extremely high sensitivity. Further, it requires an optically flat test member, therefore, limited only to plane members. However, in special circumstances, for example, photo-elastic studies where isopachics are obtained by interferometry, its use is desired. Therefore, the basic principle of interference as applied to strain measurement is first described.

Consider Fig. 5.20 in which a beam of monochromatic light from a point source S is collimated by the lens L_1. This beam is divided into two beams which travel at right angles to each other. On reflection from a flat mirror M and optically worked test member T, the beams recombine at the beam splitter and are directed in the same direction. Lens L_2 is used to

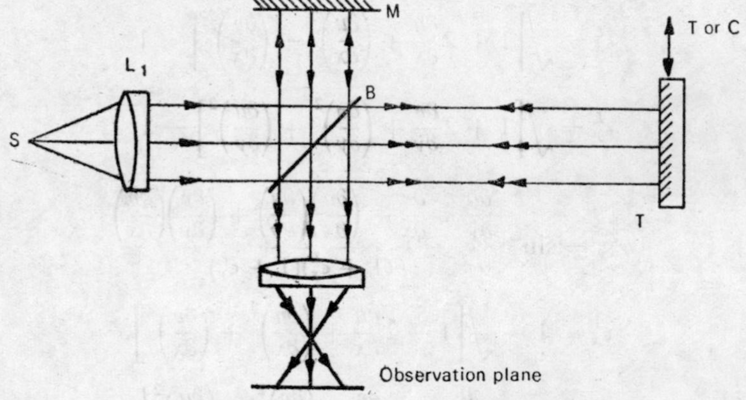

Fig. 5.20 Interferometric strain analysis

project interference fringes. Assuming that the flat mirror M and test surface T are orthogonal to each other, i.e., the reflected beams proceed in the same direction, the interference field will then be uniformly illuminated. This adjustment of the interferometer is not commonly used. Mirror M is slightly tilted which results in the straight interference fringes. The spacing of these fringes depends only on the angle of tilt and the wavelength of light. Now when the test surface is moved, the fringes move in the field of view. The interference condition is

$$2d = n\lambda, \quad n = 0, \pm 1, \pm 2, \ldots$$

Thus a movement by one fringe corresponds to a displacement d of the test mirror by $\lambda/2$. Therefore, the interferometric technique is suitable only for out of plane deformations. When the object is loaded, resulting in deformation, the fringes will take a contour of equal optical path, from which the deformations can be calculated.

Due to its extremely high sensitivity and sensitivity only to out-of-plane deformations, this technique is not often used. By certain modification the technique can be made sensitive to in-plane deformations also.

In photo-elastic studies, the difference of principal stresses $(S_1 - S_2)$ is obtained from isochromatics. However, the sum of stresses $(S_1 + S_2)$ can be obtained from interferometric measurements. Thus the magnitude of principal stresses can be obtained. One very important and often used interferometer for stress analysis is the Post's series interferometer. This instrument is shown in Fig. 5.21.

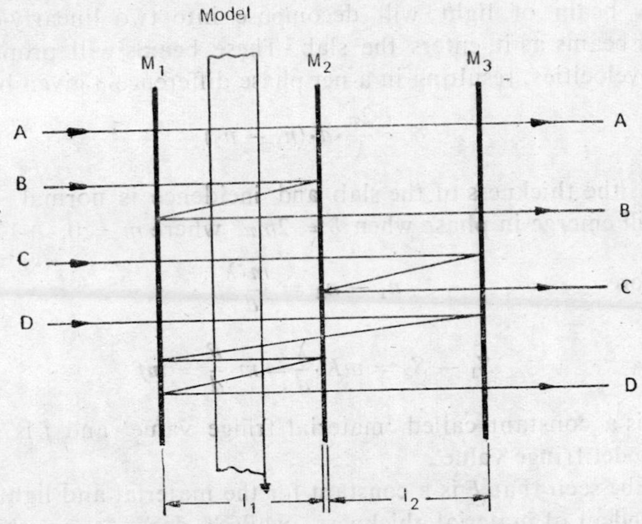

Fig. 5.21 Post's series interferometer

Post's interferometer has three partial transmitting mirrors in series. The rays *A*, *B*, *C* and *D* illustrate the transmissions and reflections taking place. Ray *A* is the directly transmitting one, while *B* and *C* suffer two reflections. Ray *D* is one of the majority of rays which undergo multiple reflections.

When the optical path l_1 is nearly equal to the path l_2, rays which traverse path similar to *B* and *C* interfere and form an interference pattern. This fringe pattern gives the difference between l_1 and l_2 at any point in the field. Superposed upon this pattern is a uniform background intensity due to all other rays transmitted through the three series mirrors.

When a transparent photo-elastic model is placed in the field between the mirrors 1 and 2 and the optical path is adjusted approximately equal to l_1, an isopachic fringe pattern representing sum of stresses $(S_1 + S_2)$ within the model is obtained.

5.11 Photo-elasticity

This technique is based on the fact that a certain class of materials when subjected to stress become birefringent. The physical properties, here dielectric constant, become direction dependent. It is found that the directions of principal stresses are along the directions of principal refractive indices. Further the difference in the magnitude of principal stresses is proportional to the refractive index difference along these axes. That is,

$$S_1 - S_2 = K(n_1 - n_2)$$

where n_1 and n_2 are refractive indices along the directions of principal stresses S_1 and S_2 and K is a constant. This relation holds good only within the elastic limit.

Consider a plane-parallel slab of a transparent material subjected to a stress. A beam of light will decompose into two linearly-orthogonal polarised beams as it enters the slab. These beams will propagate with different velocities, resulting in a net phase difference δ given by

$$\delta = \frac{2\pi}{\lambda} \cdot d \cdot (n_1 - n_2)$$

where d is the thickness of the slab and incidence is normal. These two beams will emerge in phase when $\delta = 2m\pi$, where $m = 0, \pm 1, \pm 2, \ldots$.

Therefore
$$n_1 - n_2 = \frac{m \cdot \lambda}{d}$$

Thus
$$S_1 - S_2 = mK \cdot \frac{\lambda}{d} = m \cdot \frac{F}{d} = mf$$

where F is a constant called 'material fringe value' and f is a constant called 'model fringe value'.

It may be seen that F is a constant for the material and light used, and is independent of material thickness, while f depends on the material thickness as well. F is usually employed for comparing different photo-elastic materials, while f is convenient for stress conversion in individual tests. The relation which connects $(S_1 - S_2)$ with the fringe order is called as stress optic law and is the basis of photo-elasticity. The photo-elasticity is an experimental method of stress-analysis; it is an experimental method in a sense that an optical fringe pattern is experimentally observed but the state of stress in the member is to be theoretically related with this method. This is a full field method and capable of providing a very high accuracy. As has been said that this method provides an information regarding difference of principal stresses. Therefore, to affect separation of stresses, a knowledge of S_1 or S_2 or $S_1 + S_2$ is necessary. A method to measure the sum of stresses is described later.

There are two main kinds of arrangements used in photo-elasticity: plane polariscope and circular polariscope.

Plane Polariscope

The functional elements of a standard plane polariscope and the transformations which the light undergoes when it passes through the plane polariscope are shown in Fig. 5.22 (a). The functional elements are:

(i) light source,
(ii) pair of polariser,
(iii) model, and

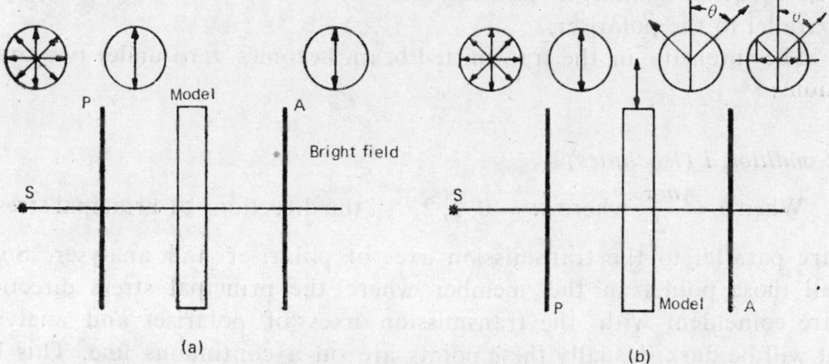

Fig. 5.22 Linear polariscope

(iv) observation screen or camera.

The lenses are used for collimating the beam and projecting the fringes at the screen. The model is placed between the polariser and analyser. Let us assume that the beam emerging from the polariser is vertically polarised (i.e., E vector vibrating vertically). This is represented as $E = E_0 e^{i\omega t}$.

If the analyser (other polariser) is crossed, i.e., its direction of transmission is perpendicular to that of the polariser, the field of view will be dark. On the other hand, if it is parallel, the field will be bright. At any other orientation, the intensity transmitted will follow $\cos^2 \theta$ law, where θ is the angle between the transmission axes of polariser and analyser. Fig. 5.22 (b) shows the same arrangement with model stressed. If one of the principal axes be tilted at an angle θ with the vertical. The amplitudes of light transmitted along the principal axes (1) and (2) are given by

$$E_1 = E_0 \cos \theta \; e^{i\omega t}$$

and
$$E_2 = E_0 \sin \theta \; e^{i(\omega t+\delta)}$$

where δ the phase difference introduced between the two beams is given by

$$\delta = \frac{2\pi}{\lambda} . d \cdot (n_1 - n_2)$$

n_1 and n_2 are principal refractive indices.

Only that component of the field, which is along the transmission direction of the analyser is allowed by the analyser. Assuming the analyser to be perpendicular with the polariser, the amplitude of the light transmitted is

$$= E_0 \cos \theta \sin \theta \; e^{i\omega t} - E_0 \sin \theta \cos \theta \; e^{i(\omega t+\delta)}$$

$$= \frac{E_0}{2} \cdot \sin 2\theta \cdot e^{i\omega t}[1 - e^{i\delta}]$$

The intensity transmitted is therefore given by

$$I(t) = I_0 \sin^2 2\theta \sin^2 \delta/2$$

An equation similar to this can also be derived when the analyser is parallel to the polariser.

The intensity in the transmitted beam becomes zero under two conditions:

Condition 1 (Isoclinics)

When $\theta = \frac{m\pi}{2}$, where $m = 0, 1, 2, \ldots$ the directions of principal stresses are parallel to the transmission axes of polariser and analyser. So at all those points in the member where the principal stress directions are coincident with the transmission axes of polariser and analyser, it will be dark. Usually these points are on a continuous line. This line goes through all points where the principal stress has the same inclination, and is called *isoclinic*. The angle θ made by the transmission axis of polariser or analyser with the reference direction (usually vertical or horizontal) is called a parameter of the concerned isoclinic. By changing θ, isoclinics of different parameters are obtained. Further it should be noted that isoclinics are independent of wavelength of light or/and of retardation (stress). Usually a white light source with a convenient stress is used to find isoclinics.

Condition 2 (Isochromatics)

The intensity of transmitted light will be zero for any value of θ, provided

$$\delta = 2m\pi \cdot m = 0, \pm 1, \pm 2, \ldots,$$

or $$d(n_1 - n_2) = m\lambda$$

Using stress optic law,

$$m = \frac{1}{f}(S_1 - S_2)$$

Therefore, in the stressed member, wherever $(S_1 - S_2)$ is such that m takes 0 or integer values, the transmitted intensity will be zero. In monochromatic light, it corresponds to continuous dark fringes. However, in white light, the appearance will be coloured. Each band corresponds to a certain value of $(S_1 - S_2)$ and, hence, is called *isochromatic*. In monochromatic light, the fringes corresponding to $m = 0, 1, 2, \ldots$ are called fringes of zero, first, second order, respectively.

Circular Polariscope

Both isoclinics and isochromatics appear together in the plane polariscope; the isoclinics are broad, dark and hide the isochromatics. Therefore, elimination of former is often desirable. This is achieved in circular polariscopes. In short, plane polariscope is used to obtain direction of principal stresses, while circular polariscope to measure difference of principal stresses. Fig. 5.23 (a) shows a schematic diagram of a circular polariscope where the assumed a configuration in which polariser and analyser are crossed with polariser's transmission axis vertical and the fast and slow axes of quarter wave plates QWP_1 and QWP_2 orthogonal to

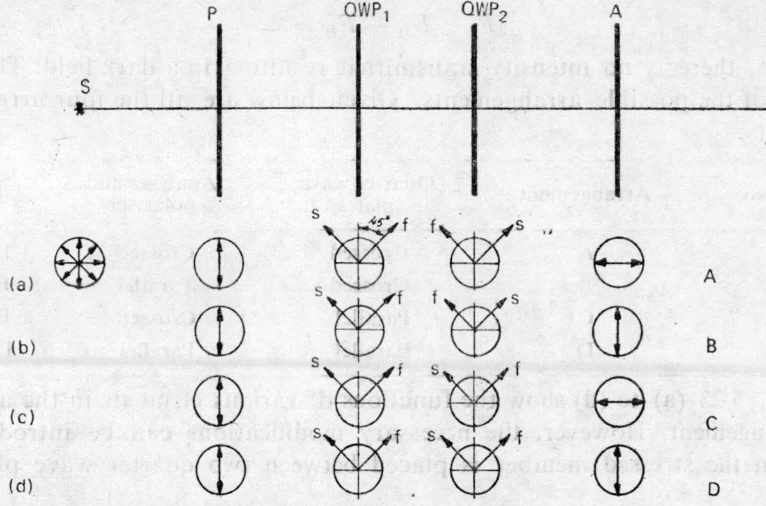

Fig. 5.23 Various arrangements in circular polariscope

each other and inclined at an angle of 45° with the transmission direction of polariser. The transfer of polarisation states through various functional elements is described below.

The electric field after the polariser is $E_v = E_0 e^{i\omega t}$. This is also the field incident on the quarter wave plate. Let the fields transmitted along fast and slow axes of QWP_1 are E_1 and E_2 respectively; then

$$E_1 = \frac{E_0}{\sqrt{2}} e^{i\omega t},$$

$$E_2 = \frac{E_0}{\sqrt{2}} e^{i(\omega t + \pi/2)}$$

These fields are incident on the quarter wave plate QWP_2 with axes orthogonal to that of QWP_1. Therefore, the fields E_3 and E_4 transmitted along slow and fast axes are

$$E_3 = \frac{E_0}{\sqrt{2}} e^{i(\omega t + \pi/2)}$$

$$E_4 = \frac{E_0}{\sqrt{2}} e^{i\omega t}$$

The fields transmitted by the analyser are those components which are parallel to the transmission axis of the analyser. Therefore,

$$E_{3t} = \frac{E_0}{2} e^{i(\omega t + \pi/2)},$$

$$E_{4t} = \frac{E_0}{2} e^{i(\omega t + \pi/2)}$$

The net amplitude transmitted by the analyser, is a vector sum of E_{3t} and E_{4t}, resulting in

$$E_t = E_{3t} - E_{4t} = 0$$

Thus, there is no intensity transmitted resulting in a dark field. This is one of the possible arrangements. Given below are all the four arrangements:

Sl. No.	Arrangement	Quarter wave plates	Analyser and polariser	Field
1.	A	Crossed	Crossed	Dark
2.	B	Crossed	Parallel	Bright
3.	C	Parallel	Crossed	Bright
4.	D	Parallel	Parallel	Dark

Figs. 5.23 (a) to (d) show the functions of various elements in the above arrangement. However, the necessary modifications can be introduced when the stressed member is placed between two quarter wave plates.

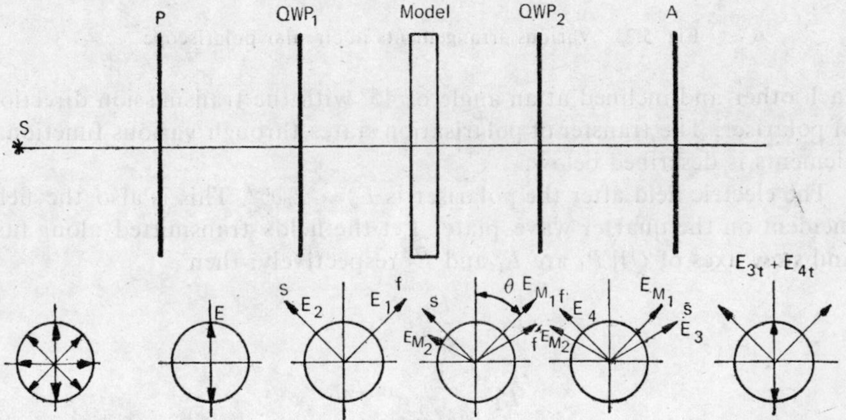

Fig. 5.24 Circular polariscope in arrangement 'B' with model stressed

Considering arrangement B and the principal axes of stressed member inclined at an angle θ with the axes of quarter wave plate as shown in Fig. 5.24, the field transmitted from the member along its axes is

$$E_{M1} = E_1 \cos\theta \cdot e^{i(\omega t + \pi/2)} - E_2 \sin\theta \cdot e^{i\omega t}$$

$$E_{M2} = (E_1 \sin\theta \cdot e^{i(\omega t + \pi/2)} + E_2 \cos\theta \cdot e^{i\omega t}) e^{i\delta}$$

or $$E_{M1} = \frac{E_0}{\sqrt{2}} (\cos\theta \cdot e_i^{(\omega t + \pi/2)} - \sin\theta \cdot e^{i\omega t})$$

$$E_{M2} = \frac{E_0}{\sqrt{2}} (\cos\theta \cdot e^{i(\omega t + \pi/2)} + \sin\theta \cdot e^{i\omega t}) e^{i\delta}$$

where $\delta = \frac{2\pi}{\lambda} \cdot d \cdot (n_1 - n_2)$.

The field transmitted by the plate QWP_2 can be obtained by taking the components of E_{M1} and E_{M2} along the axes of the quarter wave plate. Further in the *B* arrangement, the fields E_{3t} and E_{4t}, which are the components of E_3 and E_4 along the transmission axes of analyser will be added, resulting in the net field E_t to be

$$E_t = E_{3t} + E_{4t} = \frac{E_0}{2} e^{i\omega t}(1 + e^{i\delta})$$

Hence the intensity transmitted will be

$$I_t = I_0 \cos^2 (\delta/2)$$

The maxima of intensity occur when

$$\delta = 0,\ 2\pi,\ 4\pi \ldots$$

and the minima of intensity occur when

$$\delta = \pi,\ 3\pi,\ 5\pi, \ldots$$

The isochromatics of half order fringes such as 1/2, 3/2, 5/2, . . . are obtained in the bright field arrangement. Similar approach may be followed to develop necessary formulation for other arrangements.

In order to increase the accuracy, the fractional orders are to be measured. Various schemes are available, for example, compensation method, methods due to Tardy and Senarmont, etc.

EQUIPMENT

A polariscope has a light source, polarising elements, quarter wave plates, loading frame and recording/projecting arrangement. A polariscope usually has two light sources; a white light source and a monochromatic light source. For monochromatic source either sodium or mercury lamp with suitable filters is used.

In earlier polariscope, Nicol or Glan-Thomson prisms were used as polariser and analyser; the aperture thus was very much limited and hence projection arrangements were incorporated. Now the polariscopes use polaroid sheets, which are available in very large sizes. The polaroids are not very good polarisers but for photo-elastic work are quite adequate.

The quarter wave plates are usually thin sheets of mica with thickness such as to introduce a path change of $\lambda/4$. Therefore, they are wavelength sensitive. However, now quarter wave plates which are almost achromatic are available as plastic sheets where the stress birefringence is made use of for necessary phase change.

Loading frame is an element which is designed by the investigator in consideration with the shape, size, etc., of model and loading capacity. Invariably dead weight or dead weight magnified by lever is adopted in the design.

For projection, a good quality lens is used or a permanent record can be made by focusing the camera at the fringe plane.

Materials

A number of materials for model have been used; they include several types of glasses, celluloid, gelatin, rubber, cellulose nitrate, vinyls, phenol formaldehydes, kriston, CR–39, catalin 61–893, polyesters, epoxies and urethane. Of these most commonly employed are epoxy resin, castolite, catalin 61–893, Columbia resin (CR–39) and urethane. The model made of these materials can be loaded and stress field studied as a function of loads or the stresses in the model may be frozen. The frozen stress field can be studied leisurely, or for three-dimensional photo-elasticity, the model is sliced and studied.

Data Analysis

It has been shown that, using polariscope, the directions and difference of principal stresses can be experimentally determined. In order to determine the complete state of stress at any point, one should know individual stresses S_1 and S_2. Therefore, an experimental measurement of S_1, or S_2 or $S_1 + S_2$ is carried out by some alternative methods. There are a number of schemes available for this. However, only a few schemes are discussed here. It is known that the strain component perpendicular to the surface of the slab is given by

$$\epsilon_3 = \frac{S_3}{E} - \frac{\mu}{E}(S_1 + S_2)$$

where S_3 is the stress component in that direction, and μ and E are the Poisson's ratio and Young's modulus of the material of test member. Since $S_3 = 0$,

$$\epsilon_3 = -\frac{\mu}{E}(S_1 + S_2)$$

or

$$(S_1 + S_2) = -\frac{E}{\mu}\epsilon_3 = -\frac{E}{\mu}\cdot\frac{\Delta d}{d}$$

where Δd are the variations in the thickness which can be measured at various points, say by a sensitive dial gauge. This method is very cumbersome as it requires measurement at large number of points. Any interferometric method, especially the one suggested by Post, is a very useful one as it gives information about the whole field at a time. Its accuracy can be increased many times by placing the specimen between two high reflecting surfaces and only those beams which have suffered same number of reflections are allowed to interfere. This is known as fringe multiplication. Therefore, the two equations are

$$S_1 - S_2 = m\frac{F}{d} = mf$$

and
$$S_1 + S_2 = -\frac{E}{\mu} \cdot \frac{\Delta d}{d}$$

From these two equations, the state of stress at any point can be found out easily.

In another method, the model is rotated about one of the principal stress vectors at some point of interest so that the angle of incidence is ϕ. Then

$$S_1 - S_2 \cos^2 \phi = m' \frac{F}{d} = m'f$$

where m' is the order of fringe at the point determined with oblique incidence. Therefore,

$$S_2 = f\left(\frac{m - m' \cos \phi}{\cos^2 \phi - 1}\right)$$

and
$$S_1 = f\left(\frac{m \cos^2 \phi - m' \cos \phi}{\cos^2 \phi - 1}\right)$$

Other methods of separation of stresses are due to Filon, and Frocht. Electrical analogy may also be used.

It is not always necessary to make a model of the specimen for photoelastic measurement. The specimen as such can be used, provided a layer of birefringent material is applied to it and a polariscope is used in reflection mode. The sensitivity of the method is relatively low due to the thin layer of birefringent material.

5.12 Holography

Holography was invented by Gabor in 1948 to increase the resolution of an electron microscope through a two-step process involving both interference and diffraction. Later it came to be known as three-dimensional photography. Nevertheless, both are fundamentally different from each other. In photography an image of the object is made on the receiving plane by lens, mirror or their combination. This two-dimensional image when recorded does not provide any parallax and perspective for viewing. On the other hand, in holography the object wave itself is recorded; when this wave is released from the hologram, it gives a three-dimensional image of the object in space. Different perspectives of the object can be observed by changing the direction of view.

Laser is used to carry out holography successfully. It emits a highly coherent, monochromatic and intense beam. The record of object wave through the process of interference is possible because of the high coherence. Further, the record is to be made on high resolving photographic plates; specially suited for this work are 649 F Kodak, 10 E 70 and 10 E 75 Scientia Gevaert. The whole setup is mounted on a vibration free bench; no part of the setup should move by an amount greater than about 0.2 μm during the period of recording. In order to understand what holography is, how the hologram is made and reconstructed, the experimental setup is described here.

The experimental setup to record a hologram is shown in Fig. 5.25. The beam from laser is divided into two beams with the help of a beam splitter *B* and is directed in two different directions. The beam 1 is reflected by mirror M_1 and directed through a microscope objective. The microscope objective, usually 10*x*, expands the beam as to make it fill the area of photographic plate. This beam is usually termed as 'reference beam'. In order to clean the beam from the spurious noise, instead of microscope objective, a combination of microscope objective and a pin-hole is used.

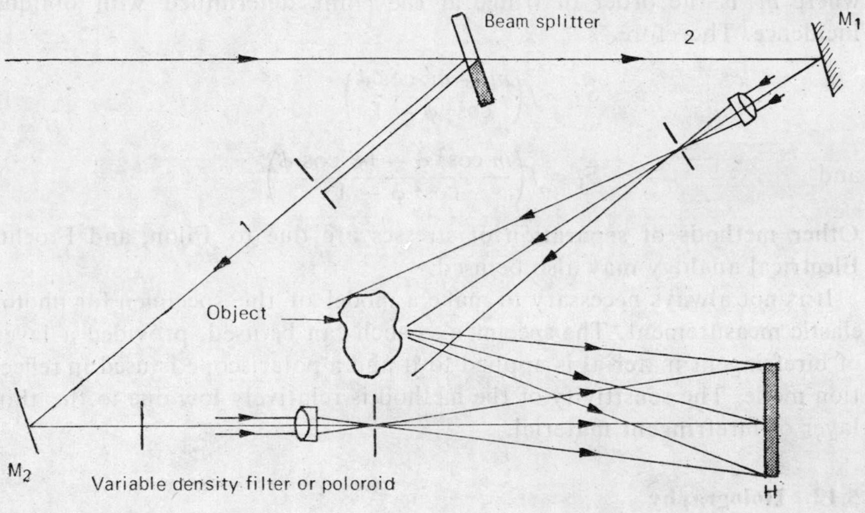

Fig. 5.25 Geometry for recording the hologram and reconstructing the object wave. For reconstruction beam 2 is blocked off and the hologram is seen through in the direction of original object.

The beam 2 after reflection from mirror M_2 is expanded by the microscope objective. This beam now fills the object. The object is usually of matt surface, otherwise a layer of Al point may be given over it. It should scatter the beam. A part of the scattered beam called object wave falls over the photographic plate. Therefore an interference pattern between the reference wave and the object wave is recorded at the photographic plate. It should be borne in mind that reference wave should be 3 to 15 times stronger than object wave at the photographic plate in order to be in the linear-region.

The photographic plate is developed, fixed and dried. The record is now called a hologram. The hologram is positioned exactly at the same place where it was during recording. On illuminating this with a reference wave and blocking beam 2, one observes in the direction of the object, an image of the object exactly at the same place where the object was, although it has long been removed. This process is known as reconstruction. It is this identity between object and its image under certain condition which has been utilised for Stress-Analysis.

Consider a situation where a hologram of the object wave is made. The hologram is repositioned accurately. On looking through the hologram, one would see an object image exactly superposed over the object, or in interference language one sees two waves, one released from the hologram and other the object wave transmitted through it, superposed exactly and hence a uniform interference field will be observed. However, if the hologram is not properly relocated, a fringe pattern will be seen, as now the two waves do not superpose exactly. But due to the wet development process, there is always a little shrinkage of the photographic record, resulting in a few interference fringes. The hologram is, therefore, so positioned as to minimise the number of fringes. This is the null position. As the object is perturbed by loading or otherwise, the object wave is perturbed which is continuously compared with the one released from the hologram. Therefore, on loading one observes an interference pattern over the object; the interference pattern changes in accordance with the perturbation or state of the stress. Therefore, this is called as 'live fringe' or 'real time' hologram interferometry. Mathematically, let $a_0e^{i\phi(r)}$ be the object wave released from hologram and $a_0e^{i\phi'(r)}$ the one from the perturbed object. The perturbation is assumed to be small so that only the phase of the object wave is changed and the amplitude remains unchanged.

The condition of interference gives

$$\phi'(r) - \phi(r) = (2m + 1)\pi$$

Therefore the bright fringe appears where the surface has been deformed by an amount which changes the phase by π. The real time hologram interferometry is capable of monitoring perturbations from the zero to a maximum value which is governed by resolving limit of the viewing system. Under such cases a number of base holograms are to be made. The main disadvantages of this are the difficulty in repositioning the hologram and the emulsion shrinkage. Both of them are removed in double-exposure holography, where two states of the object are recorded sequentially on the photographic plate. The hologram made this way generates two waves which interfere and thus a comparison between two states is made. The technique does not require repositioning of the hologram. Both the waves are equally affected by emulsion shrinkage and share the diffraction efficiency equally.

For very small deformations ($< \lambda/2$), there is no fringe formation and hence the determination of deformations of so small magnitude cannot be done by the methods described above. Subtraction techniques or background fringe techniques are used. The intensity distribution in the fringes is $\cos^2$ type. The contrast of the fringes is very good when double exposure holography is used, while in live fringe holography the contrast is poor due to strong background. The application of hologram interferometry for the study of stress field in the reflecting objects will now be described.

Consider an object with the reflecting surface shown by ABC in Fig. 5.26. Let the surface after loading be represented by $A'B'C'$; the point A

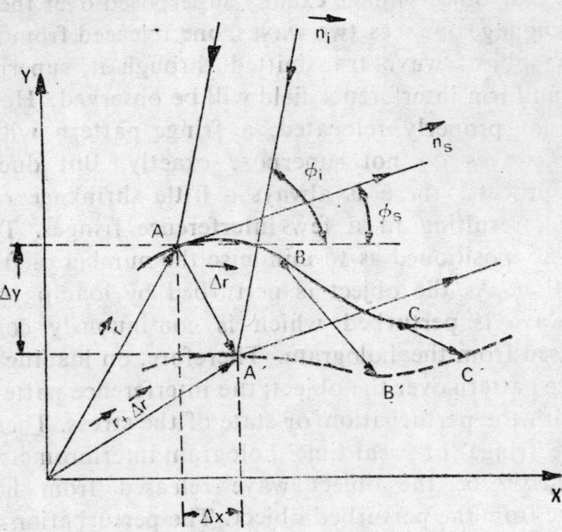

Fig. 5.26 Geometry of scattering

has moved to the point A' by an amount $\mathbf{\Delta r}$. It can be proved that the fringe formation is due to this deformation rather than the microstructure of the object which gives a speckle pattern. If the directions of illumination and observation are given by the vectors $\mathbf{n}_i$ and $\mathbf{n}_s$, then the path difference between two waves is

$$\Delta = (\mathbf{n}_i - \mathbf{n}_s) \cdot \mathbf{\Delta r}$$

The bright fringes will be formed where

$$(\mathbf{n}_i - \mathbf{n}_s) \cdot \mathbf{\Delta r} = \pm m\lambda, \quad m = 0, 1, 2, \ldots$$

Various methods have been proposed to obtain $\mathbf{\Delta r}$ from the observed interference pattern. Aleksandrov and Bonch-Bruevich suggested to use a viewing system at an aperture low enough to give ample depth of field and high enough for adequate resolving power. The viewing system is focussed on a point and the number of fringes passing over the point is counted as the viewing direction is changed. Minimum three viewing directions are used in order to get $\mathbf{\Delta r}$ as can be seen below:

Let $\mathbf{n}_{s0}$ be the initial observation direction and $\mathbf{n}_{sk}$ the direction for which k fringes have passed over the point, then

$$(\mathbf{n}_i - \mathbf{n}_{s0}) \cdot \mathbf{\Delta r} = \pm m\lambda$$

and $$(\mathbf{n}_i - \mathbf{n}_{sk}) \cdot \mathbf{\Delta r} = \pm (m \pm k)\lambda$$

Therefore, $$(\mathbf{n}_{sk} - \mathbf{n}_{s0}) \cdot \mathbf{\Delta r} = \pm k\lambda$$

In order to obtain the components of $\mathbf{\Delta r} = (\Delta x_i + \Delta y_j + \Delta z_k)$ three equations are required and hence measurements at three directions are to be carried out. It should be noted that the method as such is very sensitive to out-of-plane deformations. For the measurement of in-plane deformation, a method suggested by Ennos and Boone is to be used. Further, the fringes formed due to the interference between two waves are not localised at the surface of the object, but rather in space. This is the reason for using low aperture system for viewing in the method of Aleksandrov and Bonch-Bruevich. A method by Haines and Hildebrand requires the knowledge of the distance of localisation of fringes.

The measurement of $\mathbf{\Delta r}$ can also be accomplished by recording three holograms in three mutually perpendicular planes. In fact, this method has been applied to the measurement of components of $\mathbf{\Delta r}$ when a cantilever has been loaded. In all these methods, only the magnitude of the components of $\mathbf{\Delta r}$ can be determined. Live fringe hologram interferometry may be used to obtain the direction. The use of four holograms may also be found adequate.

A number of investigators extended the application of hologram interferometry to photo-elasticity. It is shown that hologram interferometry is capable of providing information both about sum and difference of principal stresses. A schematic of a hologram interferometer suitable for photo-elastic work is shown in Fig. 5.27. The specimen is illuminated by diffuse light by placing a diffuser in front of the object; the reference wave is a plane wave. Two holograms are made:

(i) single exposure after the model is loaded, and
(ii) double exposure, first exposure before loading and second exposure after loading the object.

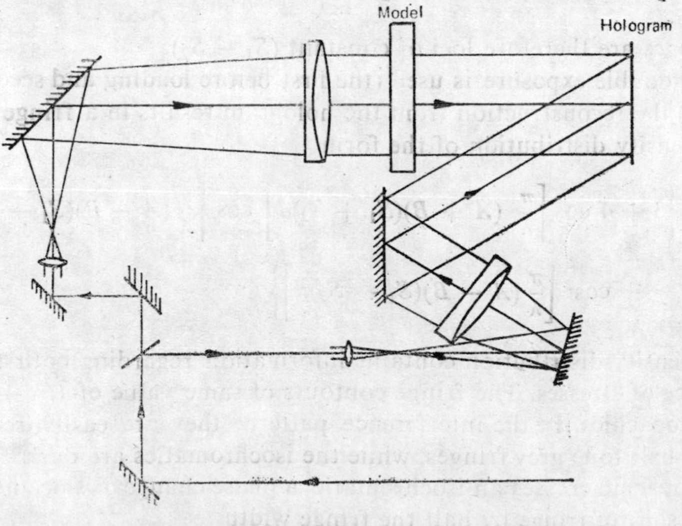

Fig. 5.27 Holo-photo-elastic interferometer

When the model is loaded, it becomes birefringent, the directions of principal refractive indices being parallel to the directions of principal stresses. The Maxwell's equations relate the principal stresses with principal refractive indices as follows:

$$n_1 - n_0 = AS_1 + BS_2$$

$$n_2 - n_0 = BS_1 + AS_2$$

where n_0 is the refractive index of the model under unstressed state, n_1, n_2 are principal refractive indices and A, B are the stress-optic coefficients to be determined by calibration. Therefore,

$$n_1 - n_2 = (A - B)(S_1 - S_2)$$

Further, the path difference introduced by the model of thickness d between these two waves is

$$\Delta = (n_1 - n_2)d = (A - B)(S_1 - S_2)d$$

When a single exposure hologram is made after the model is stressed, in fact one records interference patterns between a circularly polarised reference wave and orthogonally linear polarised beam with a path difference Δ generated in the model. On reconstruction, one observed in the object image, an interference pattern. The intensity distribution in the interference pattern is given by

$$I = I_0 \cos^2 \left[\frac{\pi}{\lambda}(A - B)(S_1 - S_2)\, d\right]$$

This, therefore, clearly gives isochromatics, whose positions are given by

$$\frac{\pi}{\lambda}(A - B)(S_1 - S_2)d = (2m + 1)\frac{\pi}{2}$$

or

$$(S_1 - S_2) = \frac{(2m + 1)}{(A - B)d}\frac{\lambda}{2}$$

The fringes are therefore loci of constant $(S_1 - S_2)$.

When double exposure is used (the first before loading and second after loading) the reconstruction from the hologram results in a fringe pattern with intensity distribution of the form

$$I = I_0\left\{3 + 4\cos\left[\frac{\pi}{\lambda}(A + B)(S_1 + S_2)d\right]\cos\left[\frac{\pi}{\lambda}(A - B)(S_1 - S_2)d\right] + \cos^2\left[\frac{\pi}{\lambda}(A - B)(S_1 - S_2)d\right]\right\}$$

This intensity distribution contains information regarding both sum and difference of stresses. The fringe contours of same value of $(S_1 + S_2)$ are called isopachics. In the interference pattern, they are easily recognised by their half tone grey fringes, while the isochromatics are dark. Further, as an isopachic crosses an isochromatic, a phase change of π occurs, resulting in a shift of fringe by half the fringe width.

FRINGE INTERPRETATION

The appearance of both isochromatics and isopachics have so far been shown and some of the experimental arrangements to observe them have also been discussed. Additional experiments are to be carried out such as calibration tests on the model material in order to determine constants A and B appearing in the equations. The calibration can be achieved by tensile test, bending test or compression test. Further on the actual model, fringe order m is to be determined. From the knowledge of A, B and m the stresses S_1 and S_2 can be obtained.

The directions of principal stresses can also be obtained by slightly modifying the arrangement to observe isoclinics as shown by Fourney.

Exercises

1. A resistance strain gauge with $R = 120$ ohms and $F = 2.0$ is placed in an equal-arm bridge in which all the resistances are equal to 120 ohms. The battery voltage is 4.0 V. Calculate the detector current in microamperes per micrometer of strain. The galvano resistance is 100 ohms.
2. In a Wheatstone bridge, leg 1 is an active gauge of advance alloy and 120 ohms resistance, leg 4 is a similar dummy gauge for temperature compensation, and legs 2 and 3 are fixed 120 ohms resistors. The maximum gauge current is to be 0.030 A.

   (i) What is the maximum permissible d.c. excitation voltage? Use this value of d.c. excitation for the remaining parts of this problem.

   (ii) If the active gauge is on a steel member, what is the bridge output voltage per 100 kg/cm² of stress?

   (iii) If temperature compensation were not used, what bridge output would be caused by the active gauge on increasing its temperature by 40°C for a steel-bonded gauge?

   What stress value would be represented by this voltage?

   Thermal expansion coefficients of steel and advance alloy are 12×10^{-6} and 25×10^{-6} cm/(cm °C), respectively. The temperature coefficient of resistance of advance alloy is 10×10^{-6} ohms/(ohm °C).

   (iv) Compute the value of shunt calibrating resistor that would give the same output as 1000 kg/cm² of stress in a steel member.
3. A single strain gauge is mounted on the centre of the aluminium bar as shown in

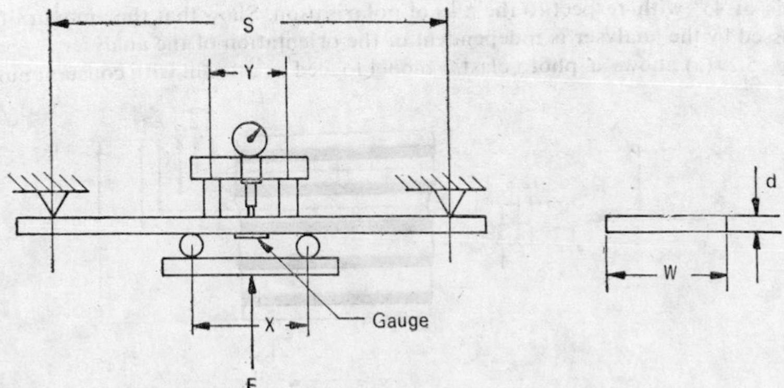

Fig. 5.28

Fig. 5.28. The bar is loaded with a constant moment section, and the curve is obtained by reading the dial indicator shown. As the bar is loaded, gauge resistance is measured by using a resistance measuring bridge. From the data given, determine the gauge factor for this gauge.

$d = 6$ mm

$w = 25$ mm

$S = 300$ mm

$X = 200$ mm

$Y = 150$ mm

R gauge	Dial reading 0.025 mm
121.3	0
120.7	4
120.2	8
119.5	12

4. A delta rosette is placed on a steel plate 2×10^6 N/m² and indicates the following strains:

$$\epsilon_1 = 395 \ \mu\text{m/m}$$
$$\epsilon_2 = 80 \ \mu\text{m/m}$$
$$\epsilon_3 = -250 \ \mu\text{m/m}$$

Calculate the principal strains and stresses, the maximum shear stress, and the orientation angle for the principal axes.

5. A rectangular rosette is placed on a steel plate and indicates the following strains:

$$\epsilon_1 = 560 \ \mu\text{m/m}$$
$$\epsilon_2 = -150 \ \mu\text{m/m}$$
$$\epsilon_3 = -475 \ \mu\text{m/m}$$

Calculate the principal strains and stresses, the maximum shear stress, and the orientation angle for the principal axes.

6. In one of the experiments with Moirè gratings, following data were available:
Thickness and width of the slab of urethane rubber = 12 mm and 125 mm; grating pitch = 8 lines/mm.

E for urethane rubber = 618

μ for urethane rubber = 0.474

A compressive load of 10 kg was applied. Calculate the pitch of Moiré fringes.

7. Two gratings of pitch of 8 lines/mm and 8.2 lines/mm, respectively, are inclined at an angle of 30° to each other. Calculate the spacing of Moiré fringes. What will be the fringe spacing if the angle between them is increased to 45°?
8. A single quarter wave plate is placed between a polariser and an analyser with its axis at 45° with respect to the axis of polarisation. Show that the amount of light passed by the analyser is independent of the orientation of the analyser.
9. Fig. 5.29 (a) shows a photo-elastic model loaded as a beam with constant moment

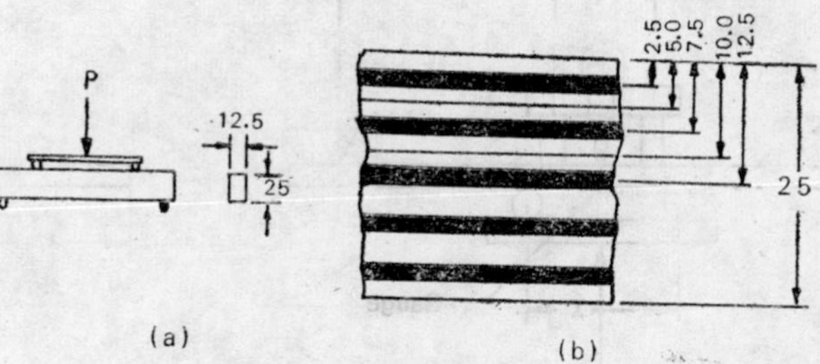

Fig. 5.29

section. If dark field is used, the fringe pattern in the constant moment section appears as shown in (b). The moment in this section is 40 cm-kg. What is the material fringe constant (F) of this material?

10. Discuss the influence of linearly polarised reference beam on the quality of hologram in the study of photo-elasticity with hologram interferometry.
11. Compare the holographic method of stress analysis with the photo-elastic method and discuss their relevant features.

6

Measurement of Pressure

6.1 Introduction

Pressure is described as force per unit area and is analogous to stress. Normally it is referred to fluids only and arises due to exchange of momentum between the molecules of the fluid and the walls of the container. The total exchange depends upon the *total number of molecules striking the wall per unit time* and the *average velocity of the molecules*. For perfect gases, using kinetic theory, it can be shown that the pressure imparted by it at the walls of the container is given by

$$p = \frac{1}{3} n \cdot m \cdot v_{\text{rms}}^2$$

where n is the molecular density, m the mass of the molecule and v_{rms} is its root mean square velocity. Further, it can be shown that

$$v_{\text{rms}} = \sqrt{\frac{3kT}{m}}$$

where k is the Boltzman constant related to the Avogadro's number N by $R = Nk$, R is the gas constant and T is the absolute temperature of the gas. Therefore, the pressure is given by

$$p = nkT$$

with $kT = \frac{1}{2}mv_{\text{rms}}^2$. So the very concept of pressure arises solely due to the kinetic nature of gas molecules. This interpretation of pressure of a perfect gas is due to kinetic theory.

The description of pressure imparted by liquids is not so simple. The kinetic theory of liquids is well developed where the interaction of molecules among themselves has been taken into account. It must, however, be noted that the pressure imparted by liquid molecules due to their momentum exchange is not a complete description; the liquid head is an additional pressure imparting process.

It is somewhat difficult to speak of the mechanism of pressure transmission at very low pressures of the order of 10^{-9} mm or lower, whereas the average size of the gas molecule is only 10^{-7} mm.

However, in practice, pressure is considered as an average physical quantity. In quite a few methods of the pressure measurements, force is measured and pressure is calculated from this by utilising the value of effective area. In other methods, indirect ways are used to obtain pressure.

6.2 Definition of pressure terms

ABSOLUTE PRESSURE. Absolute pressure refers to the absolute value of force per unit area exerted by a fluid on the walls of a container. It is the fluid pressure measured above a perfect vacuum.

ATMOSPHERIC PRESSURE. The pressure exerted by the earth's atmosphere, as commonly measured by a barometer. At sea level, its value is close to 1.013×10^5 N/m^2 absolute decreasing with altitude.

GAUGE PRESSURE. Gauge pressure indicates the difference between the absolute pressure and local atmospheric pressure.

DIFFERENTIAL PRESSURE. This is the difference between two measured pressures.

VACUUM. Vacuum represents the amount by which the atmospheric pressure exceeds the absolute pressure.

These definitions are illustrated in Fig. 6.1. The pressure at a point in a fluid is equivalent to the weight of a certain column of the fluid acting on an unit area at that point, i.e., $p = \rho gh$. This h is termed as the head.

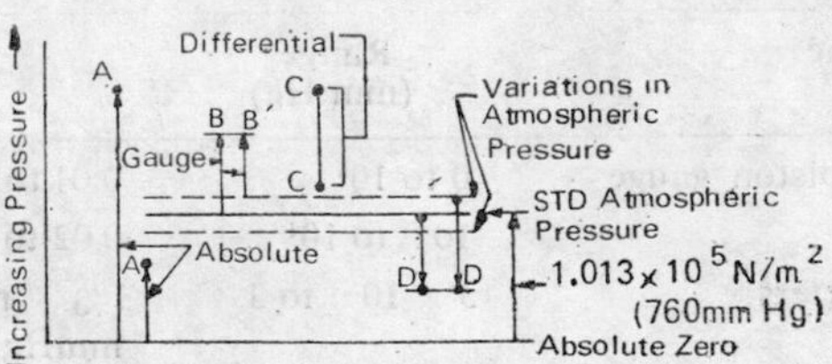

Fig. 6.1 Definition of pressure terms

6.3 Units of pressure and conversion factors

The pressure is measured either as force per unit area or a head of liquid column. It is, therefore, expressed as psi or kg/cm^2, or in mm of water or mercury. These terms are defined below:

psi pound per square inch, kg/cm^2

psia pound per square inch absolute

psig pound per square inch gauge

standard atmospheric pressure = 760 mmHg, or

$$= 14.694 \text{ psia} = 1.013\times10^5 \text{ N/m}^2$$

1 microbar = 1 dyne/cm^2, or
1 mmHg = 1333.22 microbars
1 micron (μ) = 10^{-6} m Hg = 10^{-3} mmHg
1 torr = 1 mmHg
1 mmHg = 13.619 mm water
1 kg/cm^2 = 14.52 psi
10^6 dynes/cm^2 = 14.52 psi

6.4 Methods of measuring pressure

Pressure measuring techniques can broadly be classified in the following three groups.

(i) Balancing the pressure exerted by fluid (usually mercury) column like in manometers, McLeod gauge, etc.
(ii) Measurement of elastic deformations of elements like membrane diaphragm, Bourdon tube, etc.
(iii) Measurement of electrical quantities like in Pirani and Penning gauges, Bridgman gauge, etc.

The ranges of pressure attained and the various instruments (gauges) used to measure pressure are shown in Table 6.1. The working principles of some of the common pressure measuring instruments will be discussed in this chapter.

TABLE 6.1

Type	Name	Range (mm Hg)	Uncertainty
i	Dead weight piston gauge	0 to 10^6	0.01 to 0.05% of reading
	Manometers	10^{-1} to 10^4	0.02 to 0.2% of reading
	Micromanometers	3×10^{-4} to 3	1% reading to 10^{-3} mmHg
	McLeod gauge	0 to 10^{-6}	1% of the reading
	Bourdon tube	10^2 to 10^6; up to 10^7 with certain types	up to 0.5% of full scale
	Bellows	10 to 10^4	1% of the reading
	Diaphragms (Mechanical)	0.5×10^2 to 10^5	about 0.5% of the reading
	Strain gauge bonded type diaphragm	0 to 10^6	0.25% or "better" of the reading
	Piezo-electric gauge	10^2 to 10^6	
	Conductivity (Pirani) gauge	10^{-4} to 0	
	Ionisation gauge	10^{-8} to 10^{-3}; the range can be extended down to 10^{-12}	
	Electrical resistance	10^4 to 10^9	
	Radiometer type (Knudsen gauge)	10^{-3} to 10^{-6}	

6.5 Pressure standards

Pressure standards are the basis for all pressure measurements. A gauge is always calibrated against a pressure standard and many different types are available, depending on the pressure and accuracy required. Typical standards include

(i) high accuracy 'test gauges' which themselves must be periodically checked against a more accurate standard,
(ii) 'liquid column testers' which compare the tested device against a known liquid head,
(iii) 'dead weight testers' which provide a known force by means of standard weights.

Most of these standards and other pressure sensors will now be described in detail.

6.6 Dead weight gauge

The gauge consists of an accurately machined piston which is inserted into a close fitting cylinder, both of known cross-sectional areas. This gauge is connected as shown in Fig. 6.2. A number of masses of known weight are loaded on one end of the free piston. Fluid pressure is

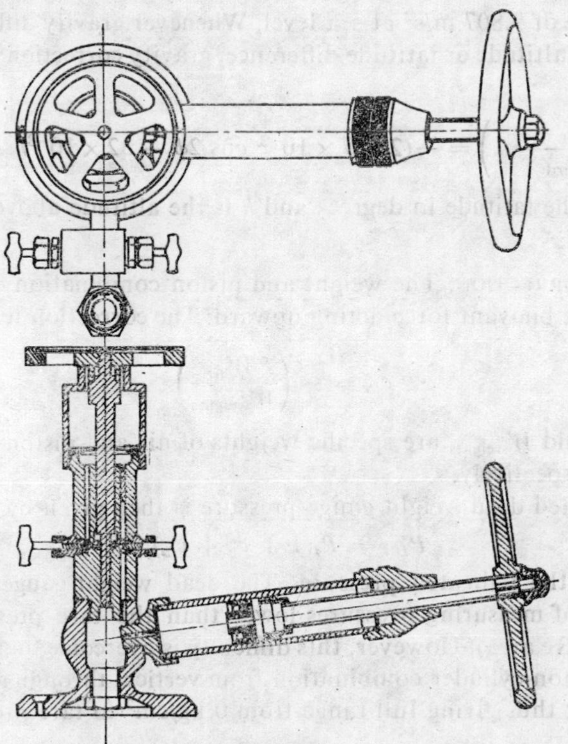

Fig. 6.2 Dead weight tester

then applied to the other end of the piston by means of some type of pump and bleed valves until enough force is developed to lift the piston-weights combination. When the piston is floating freely between limit stops, the fluid gauge pressure must be equal to the dead weight supported piston divided by the effective area of piston, i.e.

$$P_{DW} = F_E/A_E$$

where F_E the equivalent force of piston-weights combination depends on local gravity, air buoyancy, etc.; while A_E, the effective area of the piston-cylinder combination, depends on such factors as piston-cylinder clearance, pressure level, and temperature. The effective area is generally taken as mean of the cylinder and piston areas, but temperature affects this dimension. The effective area increases between 24 to 32 ppm/°C for commonly used materials, and a suitable correction can be applied for this.

There will be a fluid leakage out of the system through the piston-cylinder clearance. This fluid flow provides necessary lubrication between the two surfaces. The piston is either oscillated or rotated to reduce the friction further. Due to the leakage of fluid, the system pressure must be trimmed continuously to keep the piston-weight combination floating.

For highly accurate results, the following two corrections are to be applied.

GRAVITY CORRECTION. The weights are usually given in terms of standard gravity value of 9.807 m/s^2 at sea level. Whenever gravity differs from this value due to altitude or latitude difference, gravity correction term is given by

$$c_g = \left(\frac{g_{\text{local}}}{g_{\text{standard}}} - 1\right) = -(2.637 \times 10^{-3} \cos 2\phi + 32 \times 10^{-8}h + 5 \times 10^{-5})$$

where ϕ is the latitude in degrees and h is the altitude above sea level in meters.

BUOYANCY CORRECTION. The weight and piston combination displaces air, resulting in a buoyant force acting upward. The correction term for this is

$$c_{tb} = -\left(\frac{W_{\text{air}}}{W_{\text{weights}}}\right)$$

where W_{air} and W_{weights} are specific weights of air and piston-weight combinations, respectively.

The corrected dead weight gauge pressure is then given by

$$P_{DW} = P_I(1 + c_g + c_{tb})$$

where P_I is the indicated pressure. The dead weight gauges usually are not capable of measuring pressures lower than the tare pressure (piston weight/effective area). However, this difficulty is overcome, in some gauges, by tilting piston-cylinder combination from vertical through an accurately known angle, thus giving full range from 0 kg/cm^2 to tare pressure.

6.7 Manometers

The atmospheric pressure at NTP is 760 mm of mercury. Therefore the pressure can be measured by measuring the height of a mercury column. Usually mercury and water are preferred as manometer fluids because detailed information is available on their specific weights. Other low vapour-pressure and low density liquids like silicon oil are also used as manometer fluids.

The widely used manometer is of U-tube type, partially filled with a suitable liquid. This is widely used for measurement of fluid pressures under steady state conditions, usually capillary effects are neglected. To one end of the manometer is applied a reference pressure, say atmospheric pressure, while the other end is connected to the pressure to be measured. Consider a very general situation as shown in Fig. 6.3. The equality of pressure is taken at points 1 and 2. Therefore,

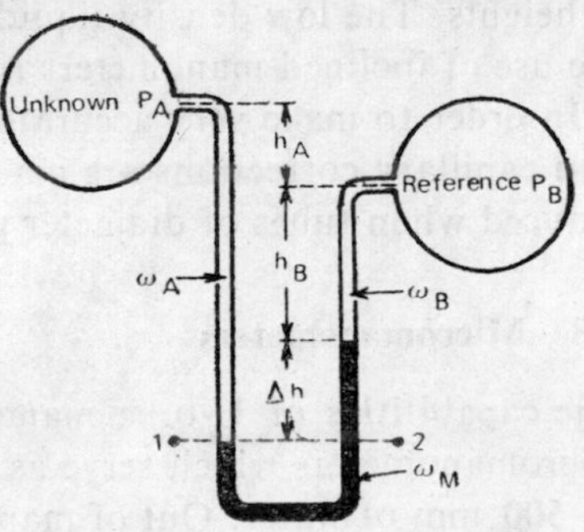

Fig. 6.3 Generalized manometer

$$P_A + W_A(h_A + h_B + \Delta h) = P_B + W_B h_B + W_M \Delta h$$

or
$$P_A - P_B = W_M \Delta h\left[1 + \frac{W_B}{W_M}\frac{h_B}{\Delta h} - \frac{W_A}{W_M}\left(1 + \frac{h_A + h_B}{\Delta h}\right)\right]$$

$$= W_M \Delta h \, c_h$$

where P_A and P_B are the unknown and reference pressures, W_A, W_B, W_M are the specific weights (corrected for temperature and gravity) of fluids through which unknown and reference pressures are transmitted and of manometer fluid, Δh, h_B, h_A are the heights as shown in Fig. 6.3. The correction factor c_h is

$$c_h = \left\{1 + \frac{W_B}{W_M}\frac{h_B}{\Delta h} - \frac{W_A}{W_M}\left[1 + \frac{h_A + h_B}{\Delta h}\right]\right\}$$

If $W_A = W_B$, i.e., similar fluids are used to transmit pressures, then

$$c_h = 1 - \frac{W_A}{W_M}\left(1 + \frac{h_A}{\Delta h}\right)$$

In addition to this, if $h_A = 0$, then

$$c_h = 1 - W_A/W_M$$

In usual practice air is used for transmitting pressure and c_h may be taken as unity, giving

$$P_A - P_B = W_M \Delta h$$

which is a familiar expression.

It should be noted that Δh is measured parallel to the gravitational force and accuracy of pressure measurement depends on the accuracy with which Δh can be measured.

Manometers can be used for the measurement of pressures at the moderate range of 10^{-1} mm to 10^4 mmHg. Measurement of pressures higher than 10^4 mmHg needs large manometer, difficult to handle, while the measurement of low pressures is difficult due to very little difference in heights. The low density liquids can be used to increase the sensitivity; the use of inclined manometers is quite common.

In order to make very accurate measurements, the temperature, gravity and capillary corrections are made. The capillary effects are considerably reduced when tubes of diameter greater than 20 mm are used.

6.8 Micromanometers

The capabilities of U-tube manometers are extended by various types of micromanometers which serve as pressure standards in the range of 0.005 to 500 mm of water. Out of many commercially available instruments a few have been described here. The main errors due to meniscus and capillary effects are minimised in all these instruments.

Prandtl Type

This U-tube type manometer consists of a reservoir of large diameter and an inclined tube with two marks, connected through a flexible tube. Two variants exist; in one the reservoir is raised to restore the liquid level between two marks, while in the other the inclined tube is moved vertically. The capillary and meniscus errors are minimised by bringing the level to a reference null position (between two marks) before the application of the pressure. After the application of pressure difference either the reservoir or the inclined tube is moved vertically by a lead screw to achieve the null position again. The motion Δh of lead screw is used to calculate pressure difference or it may be directly indicated at the dial of the lead screw.

The fundamental equation is

$$p_1 - p_2 = \frac{W_M}{W_{\text{water}}} \cdot \Delta h$$

where W_M and W_{water} are the specific weights of manometer liquid and

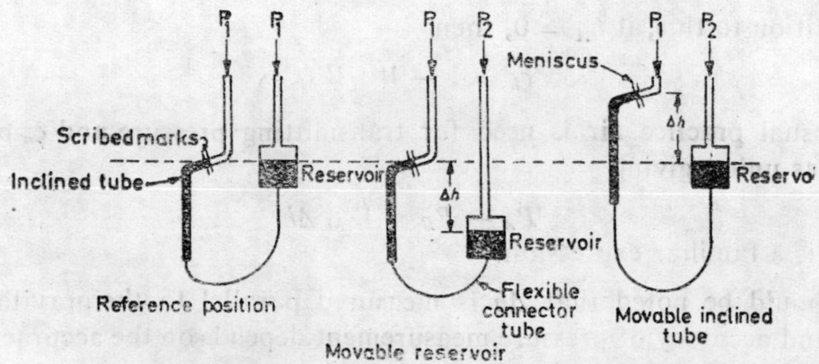

Fig. 6.4 Two variations of the Prandtl-type manometer

water, respectively. Fig. 6.4 illustrates the operation of this micromanometer.

Micrometer Type

In this type of micromanometer, meniscus and capillary effects are minimised by measuring liquid displacements with micrometer heads fitted with adjustable sharp index points located at or near the centre of large bore transparent tubes that are joined at their bases to form a U as shown in Fig. 6.5. The contact with the surface of manometer liquid may be sensed optically or electrically.

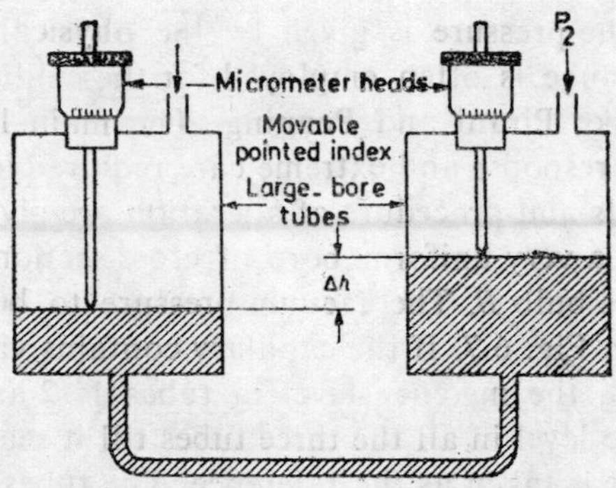

Fig. 6.5 A micrometer-type manometer

Air Micromanometer

An extremely sensitive, high response micromanometer uses air as its working fluid, and therefore avoids all capillary and meniscus effects usually encountered in liquid manometry. In this device, reference pressure is mechanically amplified by centrifugal action in a rotating disc. The disc speed is adjusted until the amplified reference pressure just balances the unknown pressure. The null position is obtained by observing the lack of movement of minute oil droplets sprayed into the glass indicator tube located between the unknown and amplified pressure (Fig. 6.6). At balance the micromanometer yields the applied pressure difference through the relation

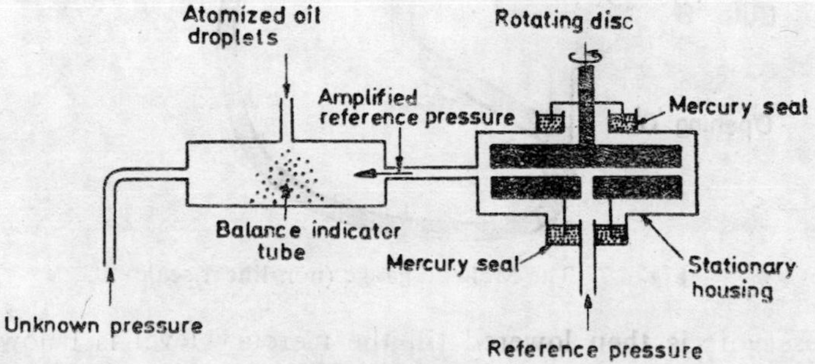

Fig. 6.6 An air-type centrifugal micro-manometer

$$p_1 - p_2 = k\rho n^2$$

where ρ is the reference air density, n is the angular speed of the disc, and k is a constant that depends on disc radius and annular clearance between the disc and housing. The measurement of pressure as small as 0.005 mm of H_2O can be made with this micromanometer with an uncertainty of 1%.

McLeod Gauge

It is a modified mercury manometer, used mainly for the measurement of vacuum pressures from 1 mm to 10^{-6} mm of Hg. It measures a differential pressure, and hence is very sensitive. Further, it measures absolute pressure because the pressure is given by the physical dimensions of the gauge. McLeod gauge is often employed for the calibration of electrical pressure gauges like Pirani and Penning. The main limitations of this gauge are its slow response and extreme care required in its handling. The construction details and procedure of operation are given below:

A capillary c of a very uniform bore of cross-sectional area A is connected with a large bulb B. The vacuum pressure to be measured is connected as shown in Fig. 6.7. If the capillary contains vacuum, then as the reservoir is raised, the mercury level in tubes 1, 2 and 3 will rise and remain at the same level in all the three tubes till it reaches the end part of the tube 1. This is taken as the reference. The tubes 1 and 2 have the same bore dimensions to avoid surface tension effects.

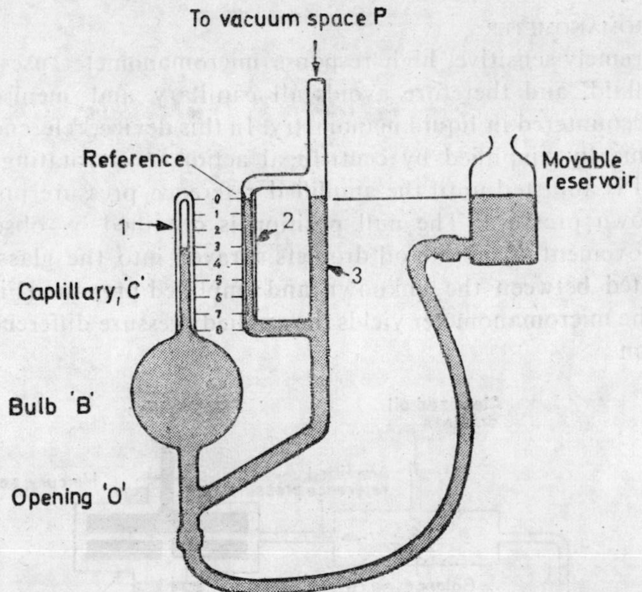

Fig. 6.7 The McLeod gauge (non-linear scale)

The reservoir is then lowered till the mercury level is below O; the pressure source is thus connected to the capillary C. Therefore, the bulb and capillary are at the pressure of the source, which is to be measured. The reservoir is raised again, thus cutting off the pressure source from the bulb. The gas in the bulb is compressed and confined to the capillary as the reservoir is moved up. When the mercury level in tube 2 has reached the reference level, the height in the capillary is measured and pressure calculated using Boyle's law. Often the capillary readings are directly calibrated in pressure.

If the height in the capillary measured from the reference mark is y, the volume of the gas enclosed is $y \cdot A$.

Let the volume of the bulb, capillary and the tube down to opening O be V_B, then

$$p \cdot V_B = A \cdot y \cdot p_c$$

where p_c is the pressure in the capillary and p is the unknown pressure. Further

$$p_c = p + W_M y$$

Thus
$$p = \frac{A W_M}{V_B - A y} \cdot y^2$$

Usually
$$A \cdot y \ll V_B$$

so
$$p = \frac{A W_M}{V_B} \cdot y^2$$

The pressure is thus obtained in terms of physical dimensions A and V_B. The tube thus can be directly calibrated in pressure.

Linearisation of pressure

The scale, as indicated by equation, is not linear but parabolic and

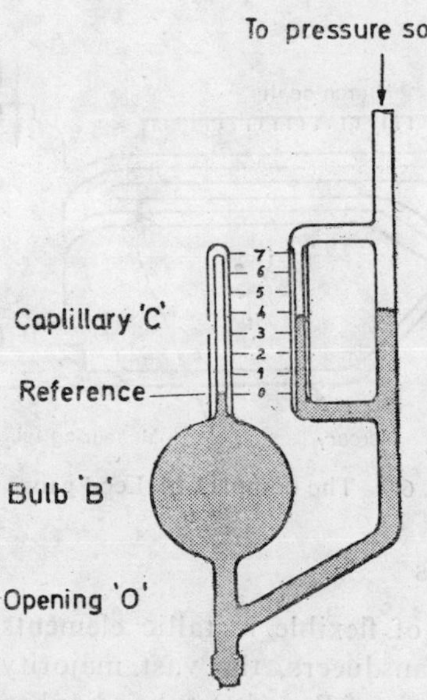

Fig. 6.8 The McLeod gauge (linear scale)

hence the sensitivity of the instrument is not constant over the whole range. The scale, however, can be linearised in this case, instead of raising mercury to the reference mark, a constant volume of gas is trapped in the capillary as shown in Fig. 6.8. Let the volume of the capillary above the fixed mark to which mercury is always raised be V_c, then

$$p \cdot V_B = p_c \cdot V_c$$

$$= (p + W_M H) V_c$$

or

$$p = \frac{V_c W_M}{V_B - V_c} \cdot H$$

$$\cong \frac{V_c W_M}{V_B} H$$

Volumes V_c and V_B are known; hence pressure can be easily computed. Note that the pressure is linearly related to height H and so the scale is linear. In using a McLeod gauge, it is important to realise that if the gas, whose pressure is monitored contains any vapour that are condensed by the compression process, pressure becomes erroneous. Except for this effect the reading of McLeod gauge is not influenced by the composition of the gas. The main drawbacks of this gauge being lack of continuous output reading and limitations on the lowest measurable pressures. When it is used to calibrate other gauges, a liquid-air trap should be used between the McLeod gauge and tested gauge to prevent passage of mercury vapour.

A very compact design of McLeod gauge is shown in Fig. 6.9. Gas at low pressure is trapped in the manometer tube by mercury as the gauge rotates about a pivot by $\pi/2$ from horizontal. Mercury compresses the gas into the top of measuring tube where final volume, expressed as initial pressure, is read.

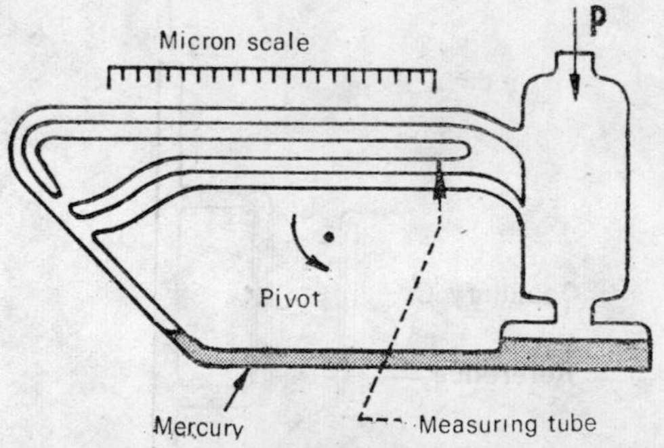

Fig. 6.9 The compact McLeod gauge

6.9 Elastic deformations

While a wide variety of flexible metallic elements might conceivably be used for pressure transducers, the vast majority of practical devices utilise one or another form of Bourdon tube, diaphragms and bellows as their sensitive elements as shown in Fig. 6.10.

The gross movement of these elements may directly actuate a pointer/scale read out through suitable linkages or gears, or the motion may be transduced to an electrical signal by one means or another. The strain gauges bonded to diaphragms are also widely used to measure local strains that are directly related to pressure.

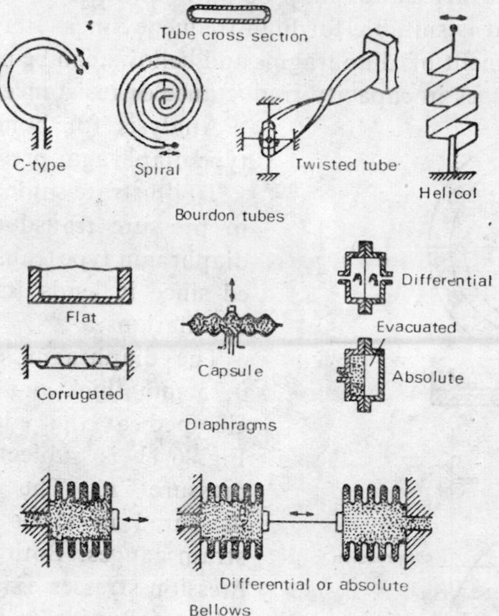

Fig. 6.10 Elastic pressure transducer

C-Type Bourdon tube is used up to pressures of 10^7 mmHg and other type versions are used up to 10^6 mmHg as the sensitivity is very high.

Flat diaphragms are widely used in electrical transducers either by sensing the central deflection with some displacement transducer or by bonding strain gauges to diaphragm surface. The pressure-deflection formula for a flat diaphragm with edge clamped is

$$p = \frac{16Et^4}{3R^4(1-\mu^2)} [Y_c/t + 0.488 (Y_c/t)^3 \ldots]$$

where Y_c is the maximum deflection; t is the thickness of the diaphragm; E, the Young's modulus; μ, the Poisson's ratio of the diaphragm material and R, the radius of diaphragm. In order to have a linear response, the second and higher order terms must be very small compared to first one. If a non-linearity of less than 5% is desired, the maximum deflection must be less than about $\frac{1}{3}$ of the diaphragm thickness. To facilitate linear response over a larger range of deflections than that imposed by the one-third thickness restriction, the diaphragms may be constructed out of corrugated discs. These types of diaphragms are most suitable for those applications where a mechanical device is used for sensing the deflection of the diaphragm.

In bellow gauge, a differential pressure causes a displacement of the bellows which may be converted to an electrical signal or undergoes a mechanical amplification to permit display of the output on an indicator dial. The bellow gauge is generally unsuitable for transient measurements because of the larger relative motion and mass involved. The diaphragm gauge, on the other hand, may be quite stiff, involves rather small displacements, and is suitable for high frequency measurements.

The displacement of diaphragms and bellows can be measured by either measuring changes in capacity, inductance or resistance.

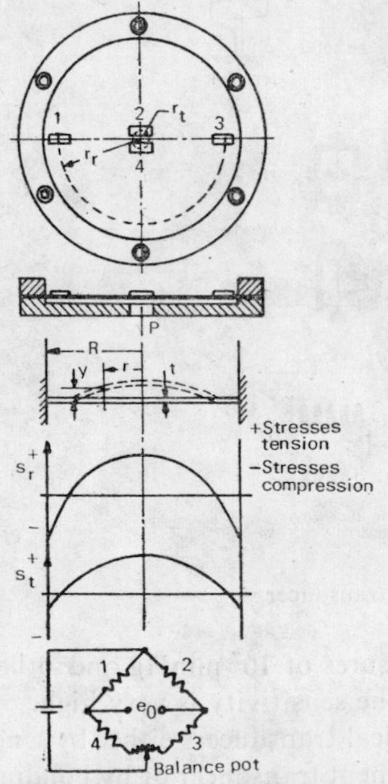

Fig. 6.11 Diaphragm type strain gauge pressure pick-up

Analysis for bonded strain gauge type diaphragm pressure gauge:

To illustrate some concepts involved in pressure transducer design, a flat diaphragm type transducer is considered since it lends itself to theoretical calculations.

The diaphragm of approximately 0.75 mm thick and 50 mm diameter, clamped at the edges as shown in Fig. 6.11 is subjected to a uniform pressure p. The displacement is measured in terms of the output of strain gauges. Both tension and compression stresses exist simultaneously.

The deflection at any point is given by

$$y = \frac{3p(1-\mu^2)(R^2-r^2)^2}{16Et^3}$$

This relation holds, provided the operation is within the linear response region. The radial stress S_r and tangential stress S_t at any point on the low pressure surface of the diaphragm, when it is subjected to uniform pressure p is given by

$$S_r = \frac{3pR^2\mu}{8t^2}\left[\left(1+\frac{1}{\mu}\right)-\left(1+\frac{3}{\mu}\right)\frac{r^2}{R^2}\right]$$

and $$S_t = \frac{3pR^2\mu}{8t^2}\left[\left(1+\frac{1}{\mu}\right)-\left(3+\frac{1}{\mu}\right)\frac{r^2}{R^2}\right]$$

The tangential stress is maximum at the centre ($r = 0$) and has a positive value

$$S_{t_{\max}} = \frac{3pR^2\mu}{8t^2}\left(1+\frac{1}{\mu}\right)$$

and the radial stress is maximum at the edge ($r = R$) and has a negative value.

$$S_{r_{\max}} = -\frac{3pR^2\mu}{8t^2}$$

The variation of S_t and S_r as a function of r/R is shown in Fig. 6.11. The above equations are accurate only for sufficiently small pressures.

The scheme of mounting strain gauges which is very satisfactory and exploits the stress situation was given by Wenk and is illustrated in Fig. 6.11. To measure the stresses and at the same time provide temperature compensation, gauges 2 and 4 are placed as close to the centre as possible and oriented to read tangential strain since it is maximum at the centre, while the gauges 1 and 3 are mounted as close to the edge as possible and oriented to read radial strain as it is maximum at that point. The laws of bridge circuitry show that the pressure effects on all the four gauges are additive and at the same time temperature effects are nullified.

The diaphragm is in a state of biaxial stress and both the radial and tangential stresses contribute to the radial and tangential strains at any point. The general biaxial stress-strain relations give

$$\epsilon_r = \frac{S_r - \mu S_t}{E}$$

and

$$\epsilon_t = \frac{S_t - \mu S_r}{E}$$

S_r and S_t the radial and tangential stress can be calculated from the parameters of the diaphragm. Using these values in the above equation, corresponding radial and tangential strain can be calculated. From the bridge sensitivity equation given below

$$e = \frac{V}{4}\left[\frac{\Delta R_1}{R_1} - \frac{\Delta R_2}{R_2} + \frac{\Delta R_3}{R_3} - \frac{\Delta R_4}{R_4}\right]$$

where e is the bridge output, V is the excitation voltage and R's and ΔR's are the resistances and changes in the resistances due to application of pressure. Further ϵ_r and ϵ_t measured by gauges 1, 3 and 2, 4 produce an additive output. Hence

$$e = \frac{VF}{4}(2\epsilon_r + 2\epsilon_t)$$

where F is the gauge factor. This gives the output in mV and can be easily computed.

Miniature pressure probes are now commercially available. A silicon diaphragm on which a Wheatstone bridge has been atomically bonded using diffusion techniques is the pressure sensing element. These are available in wide variety of ranges and possess very good frequency response. If the transducer is to be used for dynamic measurements, its natural frequency is of interest. A diaphragm has an infinite number of modes of oscillations. However, the lowest one is of interest. For a clamped-edge diaphragm vibrating in vacuum (no-fluid inertia effects) the lowest natural frequency is given by

$$\omega_n = \frac{10.21}{R}\left(\frac{gEt^2}{12W(1-\mu^2)}\right)^{1/2} \text{ per sec}$$

where W is the local specific weight of diaphragm material.

A number of factors may make the actual operating frequency value different from that predicted by the above equation:

(i) the edge clamping is never perfectly rigid; any softness tends to lower ω_n.

(ii) wrinkling tends to stiffen the diaphragm and thus raise ω_n,

(iii) if the diaphragm is used to measure liquid pressures, inertia of liquid tends to lower ω_n,

(iv) when it is used with gases, the volume of gas trapped behind the diaphragm may act as stiffening spring, thus raising ω_n.

6.10 Electrical methods

The pressure transducers which provide an output as an electrical signal fall under this group. A resistance pressure transducer which is used for the measurement of very high pressures is discussed first.

Measurement of very high pressures

The high pressure range has been defined as beginning at 10^6 mm Hg and extends to 10^8 mmHg. Very high pressures may be measured by electrical resistance gauges, which make use of the resistance change brought about by the direct application of pressure to the electrical conductor itself. The sensing element consists of a loosely wound coil of relatively fine wire. When pressure is applied, the bulk compression effect results in an electrical resistance change that may be calibrated in terms of applied pressure.

The resistance of a wire is given by

$$R = \rho L/A = \rho L/CD^2$$

where C is a proportionality constant, its value being $\pi/4$ for circular cross section; other symbols having their usual meaning.

On differentiation,

$$\frac{dR}{R} = \frac{d\rho}{\rho} + \frac{dL}{L} - \frac{2dD}{D}$$

In the case of freely suspended wire in the pressure medium, the wire is subjected to biaxial stress condition only, because the ends, in providing electrical continuity, will generally not be subjected to pressure. Thus $S_x = S_y = -p$ and $S_z = 0$, where S_x, S_y, S_z are stresses along x, y and z directions. Hence, the strains along x, y, z directions are given by

$$\epsilon_x = \epsilon_y = \frac{dD}{D} = -\frac{p}{E}(1-\mu)$$

$$\epsilon_z = \frac{dL}{L} = \frac{2\mu p}{E}$$

Thus

$$\frac{dR}{R} = 2\frac{p}{E} + \frac{d\rho}{\rho}$$

If specific resistance ρ does not depend on pressure, $\frac{d\rho}{\rho}$ can be neglected and hence

$$\frac{dR}{R} = 2\frac{p}{E}$$

or

$$R = R_0(1 + bp)$$

where $b = 2/E$ is called the pressure coefficient. The resistance varies linearly with pressure.

The pressure transducer based on this principle is called Bridgeman gauge. Fig. 6.12 illustrates a typical gauge. The sensitive wire is wound

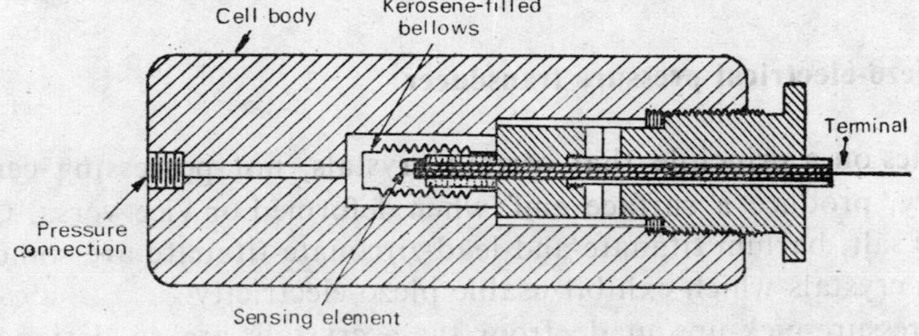

Fig. 6.12 The Bridgeman gauge

in a loose coil, one end of which is ground to the cell body and other end brought out through a suitable insulator. The total resistance of coil is around 100 ohms. The coil is enclosed in a flexible kerosene-filled bellows which transmit the pressure to the coil. The resistance change is sensed by conventional Wheatstone bridge methods. Two metals are commonly used for resistance gauges—manganin and an alloy of gold. Both metals provide linear output; their pertinent characteristics are given in Table 6.2.

TABLE 6.2

Characteristics	Manganin 84 Cu, 12 Mn 4 Ni	Gold-chrome 97.9 Au, 2.1 Cr
Pressure sensitivity or pressure coefficient (ohm/ohm)/kg/cm²	24.1×10^{-7}	9.55×10^{-7}
Temperature sensitivity (ohm/ohm)/°C	3.06×10^{-5}	1.44×10^{-6}
Resistivity ohm cm	45×10^{-6}	2.4×10^{-6}

Although the pressure sensitivity of gold-chrome is lower than manganin, it is preferred in many cases because of its much smaller temperature error. This is particularly significant since the kerosene used in bellows will experience a transient temperature change when sudden pressure changes occur, because of adiabatic compression or expansion.

Since the variation of resistance is associated with elastic movements within the wire, it occurs within the time required for sound wave to travel the wire radius. For a typical wire of 0.025 mm radius, this is of the order of 10^{-8} s. The dynamic response is therefore very good; the wire resistance changes with the application of pressure almost instantaneously. However, the accompanying temperature change will cause a transient error if temperature sensitivity is too high. Gauges of this type are commercially available with scale up to 15000 kg/cm^2 and inaccuracy of 0.1 to 1.5%. They have been used to measure pressures as high as 100,000 atmospheres.

6.11 Piezo-electrical pressure transducer

It operates on a principle that certain crystals, not possessing centre of symmetry, produce a surface emf when deformed or vice versa. Quartz, Rochelle salt, barium titanate and lead-zirconate-titanate are some of the common crystals which exhibit usable piezo-electricity.

The pressure pick-ups made from these crystals are so designed that they show maximum piezo-electric response along the desired direction with no or very little response along the other direction. As an example, a quartz x-cut crystal of 2.5 mm thickness would have a sensitivity of about 10 volts/kg/cm^2. Such high sensitivity is typical of piezo-electric transducers. The use of piezo-electric transducer elements is primarily limited to the dynamic measurements. Hence, these elements are extensively used in sound pressure instrumentation, in accelerometers and vibration pick-ups. Some commercially available systems using quartz transducers (very high leakage resistance) and electrometer input amplifiers (very high input impedance) achieve an effective total resistance of 10^{14} ohms which gives sufficiently slow leakage to allow static measurements.

Although the emf developed by piezo-electric transducers may be proportional to pressure, it is nonetheless difficult to calibrate them by normal static procedures. An attractive technique called 'electrocalibration' has been developed in which piezo-electric transducer is excited by an electric field rather than by an actual physical pressure to obtain calibration.

The dynamic response of piezo-electric pressure transducers is very good; they possess very little internal damping ($\zeta \simeq 0.007$) and very high response frequency of the order of 100 kHz.

6.12 Measurement of vacuum pressures

Thermal Conductivity Gauge

At low pressures, kinetic theory of gases predicts a linear relationship between pressure and thermal conductivity. Conductivity of gas is measured by measuring the temperature of the heated filament kept in the container with gas. The temperature of the heated wire carrying the current will depend on two factors:

(i) the magnitude of current, and
(ii) the heat loss, both conductive and radiative.

The radiation losses can be minimised by using materials of low emissivity. The conduction loss depends on the composition of gas and hence calibration stays valid only for a particular composition.

The most common type of conductivity gauges are thermocouple, resistance (Pirani) and thermistor type and differ only in the way the temperature is measured. In thermocouple gauge, the temperature of the filament is measured by a thermocouple welded to it.

In resistance type gauges, the temperature is measured indirectly. As the temperature of wire changes, its resistance changes too. Thus it is the resistance change which is a measure of temperature variation. The pressure is therefore measured in terms of resistance changes. The resistance change is measured by using conventional Wheatstone circuitry. In order to minimise/compensate the effect of ambient temperature variations, a dummy gauge or compensating cell is used as shown in Fig. 6.13.

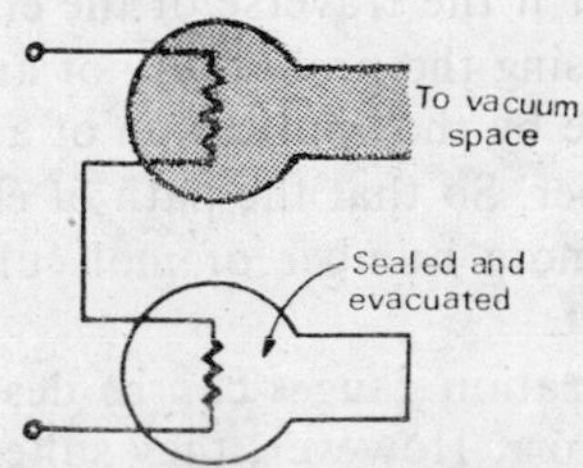

Fig. 6.13 Pirani gauge arrangements to compensate for changes in ambient temperature

In thermistor type gauges, the resistance element is thermistor—its resistance changes very rapidly with temperature and hence is very sensitive. Therefore, the compensation for ambient temperature variations is very important.

Ionisation Gauge

An electron in an electric field can be accelerated such that it can ionise the molecule of a gas when it collides, leaving a positively charged ion and a negatively charged electron. The pressure of gas is proportional to density of molecules and hence ion current is proportional to the pressure.

Figure 6.14 shows basic elements of an ionisation gauge. It is very similar to an ordinary triode electronic tube. It possesses a heated filament, a positively biased grid and a negatively biased plate, in an envelope connected to the pressure source. The electrons emitted from the filament are accelerated by the grid; they collide with gas molecules and ionise them. The ions are attracted to the plate, causing a flow of current i_i in the external circuit which is proportional to pressure. The electron current i_e which flows in the grid circuit is not affected by the secondary electrons due to ionisation and remains practically constant. The sensitivity of gauge is defined as

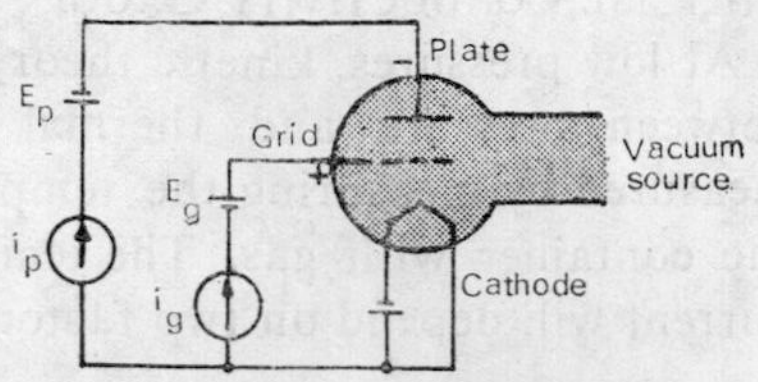

Fig. 6.14 Schematic of an ionisation gauge

$$S = \frac{i_i}{p}$$

where i_i = ion current (output) and

p = pressure (input).

As the pressure drops, ionisation current decreases as there are less number of molecules to be ionised. Ionisation current can be increased further if the traverse of the electrons from the filament is increased, thus increasing the probability of an electron colliding with a molecule. This is done by the application of a magnetic field perpendicular to the plane of paper. So that the path of electrons is helical and hence the electrons meet more number of molecules in the path, causing higher ionisation current.

Ionisation gauges can be designed to measure vacuum pressures up to 10^{-12} torr. However, they suffer from the following two disadvantages:

(i) excessive pressure (1 to 2 μm) will cause rapid oxidation of filament and thus shorten its life, and

(ii) the electron bombardment is a function of filament temperature, thus requiring a careful control of the filament temperature.

These disadvantages are eliminated in 'alphatron' where a radium source is used to ionise the gas.

Knudsen Gauge

This is another kind of gauge, which provides an absolute measure of pressure. Like McLeod gauge it shares the desirable feature of composition insensitivity but for the variation of accommodation coefficient from one gas to another (the accommodation coefficient is a measure of the extent to which rebounding molecule has attained the temperature of the heater surface). Further it is capable of giving continuous output readings.

Two vanes V along with the mirror are suspended by a very fine filament—the restoring force is provided by the torsion in the filament (Fig. 6.15). Near the vanes are two heater plates H, maintained at temperature T and are so arranged that one heater is in front of one vane and the other behind the second vane. The separation between the plate and vane is less than the mean free path of the surrounding gas, which is at temperature T_g. The vanes are at the gas temperature.

The molecules striking from the heater side impart a higher momentum due to being at higher temperature than from the other side. Thus there is a net momentum imparted to the vanes, causing them to rotate about the suspension. The rotation is monitored by a light pointer. The total momentum change depends on the molecular density, which in turn is related to the pressure and temperature of the gas. An expression for the pressure may be derived in terms of the measured force F and temperatures T and T_g as follows.

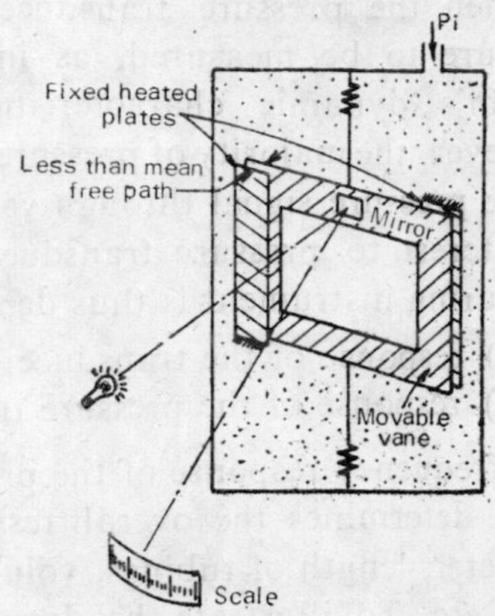

Fig. 6.15 The Knudsen gauge

The velocity of molecules at T_g is $v_0 = \sqrt{3kT_g/m}$, and the velocity of molecules at T is $v_1 = \sqrt{3kT/m}$, where m is the mass of a single molecule and k is the Boltzmann constant. The net momentum transfer per molecule is

$$m(v_1 - v_0) = \sqrt{3km}\,(\sqrt{T} - \sqrt{T_g})$$

Let the molecular density be n; therefore $n/6$ molecules are moving in one direction and their velocity is v_0. Thus the number of molecules which hit the vane per second is $(n/6)v_0$. Therefore, the rate of momentum transfer to the vane is

$$\frac{n}{6}\,v_0 \cdot \sqrt{3km}(\sqrt{T} - \sqrt{T_g}) = F$$

or

$$F = \frac{nkT_g}{2}\left(\sqrt{\frac{T}{T_g}} - 1\right)$$

Writing $p = nkT_g$,

$$F = \frac{p}{2}\left(\sqrt{\frac{T}{T_g}} - 1\right)$$

For smaller temperature difference $(T - T_g) \ll T$; thus

$$\sqrt{\frac{T}{T_g}} = \sqrt{1 + \frac{T - T_g}{T_g}} = 1 + \frac{T - T_g}{2T_g}$$

giving

$$F = \frac{p}{4} \cdot \frac{T - T_g}{T_g} = K \cdot \theta$$

where θ is the rotation and K is a torsional constant. The pressure is thus known in terms of measurable quantities.

The Knudsen gauge is a suitable device for calibrating other pressure gauges between 10^{-2} to 10^{-8} torr.

Dynamics of pressure transducers

When the pressure transducer can be directly exposed to the fluid pressure to be measured, as in certain piezo-electric transducers, the system's dynamic characteristics are those of the transducer itself. However, the majority of pressure measurements involve fluid transmission of the pressure signal through various tubes and chambers from the point of interest to pressure transducer. The transient response of pressure-measuring instrument is thus dependent on two facts:

(i) response of the transducer, and
(ii) response of the pressure transmitting fluid and tubing, etc.

The frequency response of the pressure-transmitting fluid and tubing, etc., often determines the overall response. Further the response depends on diameter, length of tubings, volume of pressure chambers, pressure difference, etc. To illustrate the dynamic response of pressure system with a very simple analysis which is restricted to small pressure differences, steady-state laminar flow conditions are assumed to exist.

Initially, assume that $p_i = p_m = p_0$ where p_i and p_m are desired measured pressure over and above p_0. p_i is assumed to vary in some fashion and it is desired to know how p_m changes. Consider the pressure transmission through a tubing of length L and diameter $2r$ to the chamber of volume V as shown in Fig. 6.16. The following forces acting in the system are considered:

(i) the force of magnitude $\pi r^2 p_i$ due to pressure p_i,
(ii) the viscous force due to wall shearing stress is $8\pi\mu L\dot{x}$, where x is the displacement of slug of gas in the tube, and μ is the fluid viscosity.

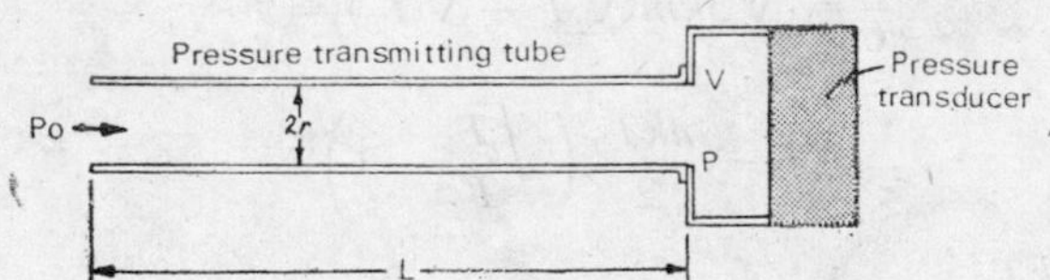

Fig. 6.16. Schematic of pressure transmitting system

If the slug of gas moves into volume V an amount of x the pressure p_m will increase. Assuming that compression occurs under adiabatic conditions, the adiabatic bulk modulus E_a of gas is

$$E_a = -\frac{dp}{(dv/V)} = \gamma p$$

The volume change is given by $dV = \pi r^2 x$. Thus the excess pressure $p_m = \pi r^2 \times E_a/V$. The force due to this pressure is

$$\pi^2 r^4 \times E_a/V$$

Hence the force balance equation is

$$\pi r^2 p_i - 8\pi\mu L\dot{x} - \pi^2 r^4 \times \frac{E_a}{V} = \frac{4}{3}\pi^2 r^2 \rho L\ddot{x}$$

where effective fluid mass is taken as 4/3 of the actual mass. Since $p_m = \pi r^2 \times \frac{E_a}{V}$,

$$\frac{4VL\rho}{3E_a r^2}\ddot{p}_m + \frac{8\mu LV}{\pi E_a r^4}\dot{p}_m + p_m = p_i$$

This is the second order equation. If an input wave of frequency ω is impressed over this system, its frequency response is given by

$$\left|\frac{p_m}{p_i}\right| = \frac{1}{\{[1-(\omega/\omega_n)^2]^2 + 4\zeta^2(\omega/\omega_n)^2\}^{1/2}}$$

where the natural frequency ω_n and damping coefficient ζ are defined as

$$\omega_n = \sqrt{\frac{3\pi r^2 c^2}{4LV}}$$

and

$$\zeta = \frac{2\mu}{pcr^2} \cdot \sqrt{\frac{3LV}{\pi}}$$

where $c = \sqrt{(E_a/\rho)}$ the velocity of sound in the medium. The phase angle for the pressure signal is

$$\phi = \tan^{-1}\frac{-2\zeta(\omega/\omega_n)}{1-(\omega/\omega_n)^2}$$

When the chamber volume is relatively large, and the transmission tube is of very small diameter, the system characteristics will approach those of a first order system i.e.

$$\left|\frac{p_m}{p_i}\right| = \frac{1}{[1 + 4\zeta^2(\omega/\omega_n)^2]^{1/2}}$$

In this case the chamber is analogous to a capacitor, and tube analogous to a resistor. Such a system, acoustical filter, can be designed to attenuate frequencies above any given value.

6.13 Considerations for pressure gauge calibration

Static Calibration

The familiar dead weight gauge may be used to provide reference pressures with which transducer outputs may be compared. The gauges of this type are useful up to 10^5 mmHg, and by the use of special designs the limit may be extended to 10^6 mmHg. The linearity of Bridgeman gauge may be used to extend the range beyond this. For medium pressures, the precision mercury columns (manometers) are used as calibration standards.

For vacuum pressures in the range of 10^{-1} to 10^{-3} torr the McLeod gauge is considered as standard. For pressures below 10^{-3} torr, a pressure dividing technique using flow through a succession of accurate orifices to relate the low down-stream pressure to a higher up-stream pressure which is accurately measured by McLeod gauge is presently in use. The technique can be further improved by substituting ionisation gauge with McLeod gauge. This must be calibrated against McLeod gauge at one point ($\leqslant 9 \times 10^{-2}$ mmHg) and its known linearity is then used to extend the range to much lower pressures up to 10^{-7} torr.

Dynamic Calibration

The dynamical response is obtained either by impulse step or frequency response tests, the step function tests being perhaps the most common.

Step function tests are used for systems whose frequencies are not greater than 1 kHz. The step input is obtained by bursting a thin diaphragm subjected to gas pressure. A general rule for step testing is that the rise time of the step function must be less than one-fourth of the natural period of the system to be tested if it is to excite natural oscillation. For pressure pick-ups with frequency greater than 1 kHz, shock tube is used to provide step input.

Exercises

1. Prove that the transient response of an idealised manometer is described by a second order equation. Express its natural frequency and damping coefficient in terms of the parameters of the manometer.
2. A diaphragm pressure gauge is to be constructed of phosphor bronze ($E = 1.12 \times 10^{11}$ N/m^2, $\mu = 0.3$) 60 mm in diameter and is to be designed to measure a maximum pressure of 15 kg/cm^2. Calculate the thickness of the gauge required so that the maximum deflection is one-third this thickness. Calculate the natural frequency of this diaphragm.
3. Design a pressure pick-up and bridge circuit to meet the following requirements:

   Maximum pressure = 10 kg/cm^2
   Natural frequency in vacuum = 10 Hz minimum
   Maximum non-linearity = 3%
   Full scale output = 10 mV minimum
   Diaphragm material is stainless steel ($E = 2 \times 10^{11}$ N/m^2)

   Strain gauges of 300 Ω resistance, gauge factor of 2.0 and size 8 mm by 8 mm are to be used.
4. A pressure pick-up has the following characteristics:

   $R = 80$ mm, $E = 2 \times 10^{11}$ N/m^2; gauge resistance = 120 ohms
   $r_r = 60$ mm, $\mu = 0.26$; gauge factor = 2.0
   $r_t = 10$ mm, $W_M = 0.01$ kg/cm^3; battery voltage = 5.0 V
   $t = 1$ mm.

   (a) Calculate the sensitivity in mV/kg/cm^2.
   (b) What is the natural frequency in vacuum?
   (c) What is the maximum allowable pressure for 2% non-linearity? What is the voltage output at this point?
5. A well type manometer uses a special bromide fluid having a specific gravity of 2.95. The well has a diameter of 75 mm and the tube has a diameter of 5 mm. The manometer is to be used to measure differential pressure in a water flow system. The scale placed along the tube has no correction factor for the area ratio of the

manometer. Calculate the value of this factor which may be multiplied by the manometer reading in mm to find the pressure differential in kgf per sq. cm.

6. Calculate the resistance change of 100 ohm coil of manganin and gold-chrome for 3.5×10^8 N/m² pressure and 40°C temperature changes.

7. It is proposed to construct a Bridgeman gauge for the purposes of measuring short-duration pressure pulses in gas. The pulses will be as large as 3.5×10^8 to 7.0×10^8 N/m². The pressure pulse will be accompanied by a temperature pulse of the order of 500°C. The active element is to be 25 μm diameter manganin wire, 25 mm long, in a channel 12 mm in diameter by 10 mm deep as shown in Fig. 6.17. (a) Design a Wheatstone bridge circuit for this unit and estimate the system sensitivity. (b) Estimate the frequency response with (i) the system as shown, and (ii) the cavity filled with grease (silicon).

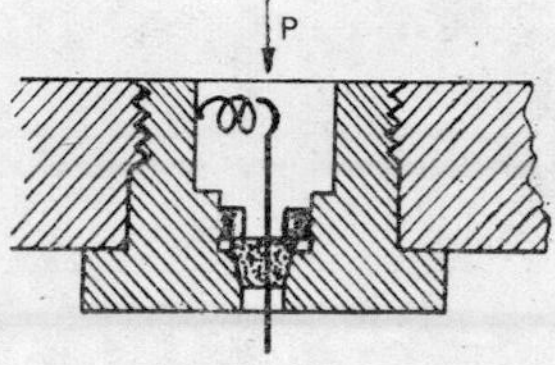

Fig. 6.17

8. A McLeod gauge has $V_s = 100$ cm³ and a capillary diameter of 1 mm. Calculate the pressure indicated by a reading of 30 mm. That error would result in the measurement if the volume of capillary is dropped in comparison with the volume of bulb.

9. A Knudsen gauge is to be designed to operate at a maximum pressure of 1.0 μm. For this application the spacing of the vane and plate is to be less than 0.3 mean free path at this pressure. Calculate the force on the vanes at pressures of 1.2 μm and 0.02 μm, when the gas temperature is 293°K and the temperature difference is 50°K.

10. A very sensitive pressure transducer is to be used to measure small variations in pressure. In order that the unit is not to be damaged by relatively rapid high pressure pulses, an acoustical filter is employed. The dimensions are given in Fig. 6.18. For air at room temperature, estimate the response of the device (assuming that mechanical response is not limiting). Note that for small pressure changes, the gas flow in the tube can be approximated by incompressible flow.

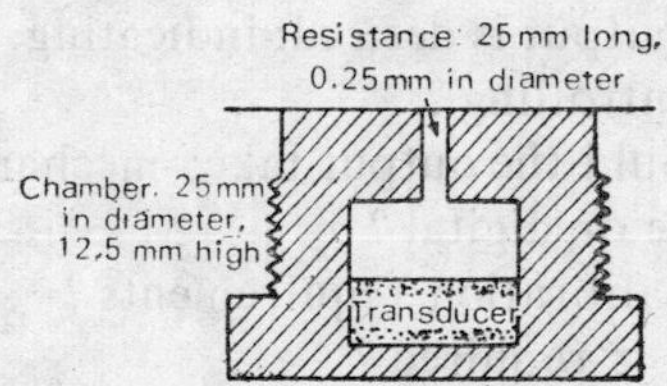

Fig. 6.18

7

Flow Measurement and Flow Visualisation

7.1 Introduction

The measurement of fluid flow is important in applications ranging from measurement of blood flow rates in human artery to the measurement of the flow of liquid oxygen in a rocket. The degree of difficulty of measurement is affected by environmental conditions and requirements of measurement. The physical and chemical properties of the fluid make measurements more difficult. Two-phase flow, with dramatic changes in volume and energy, probably, presents the greatest problem but two-component flow (gas/liquid and slurries) is not so difficult, although the distribution of mass and kinetic energy within the flow is non-uniform. The problems associated with the flow measurement are so varied and complex that everyone concerned with its measurement must be clear about the following points in advance:

(i) Whether volume flow or mass flow is to be measured?
(ii) Whether both the components of a (two-component) flow are to be measured?
(iii) What kind of output is desired: indicating, recording, integrating, batching or controlling?
(iv) What form should the output take: mechanical, pneumatic, electrical, analogue or digital?
(v) What are the calibration requirements?
(vi) What accuracy is required?
(vii) How important is reliability and what life expectancy is required?
(viii) What are the other parameters like linearity, rangeability, repeatability, dynamic response, absolute accuracy, etc.?

The basic principles of flow measurement are developed here and no attempt has been made to discuss the generalised approach to it. Out of wide variety of devices used, only typical ones will be discussed under various categories.

TABLE 7.1 Performance characteristics of flow me[illegible]

Scale	Max. viscosity (centistokes)	Temp. range °C	Max. pressure kg/cm²	Useful range
Sq. root, rate	4000	270 to 1000	420	[illegible] : 1
Sq. root, rate	4000	−50 to 800	100	[illegible] : 1
Sq. root, rate	4000	−50 to 800	100	[illegible] : 1
Sq. root, rate	1000	−50 to 1000	350	[illegible] : 1
Linear, rate	100	−40 to 200	15	[illegible] : 1
Linear, rate	100	−180 to 850	350	[illegible] : 1
Linear, total	1000	−20 to 120	15	[illegible] 20 : 1
Linear, total	2000	−20 to 150	200	[illegible] 20 : 1
Linear, total	2000	−20 to 150	10	[illegible]
Linear, total	all gas	−35 to 80	7	[illegible]
Linear, total	30	−270 to 540	1050	[illegible]
Linear, rate	1000	−130 to 180	40	[illegible]
Linear, rate	1000	−180 to 315	105	[illegible]
Linear, total, rate	20	−20 to 70	100	[illegible]
Linear, total, rate	20	−250 to 450	350	[illegible]
Logarithmic rate	all gas	−20 to 40	40	[illegible]

Line size mm	Installation recommendations	Typical power requirement	Relative cost
[illegible] and larger	Straight pipe 10 dia. upstream 3 downstream	None	1.0
25–600 and larger	Straight pipe 10 dia. upstream 3 downstream	None	[illegible]
25–600 and larger	Straight pipe 10 dia. upstream 3 downstream	None	1.5
[illegible]	Vertical or horizontal	115 V, 60 Hz, 20 W	[illegible]
[illegible]–100	Vertical only	None	1.0
[illegible]–300	Vertical only	None	1.2
[illegible]	Horizontal recommended	None	[illegible]
[illegible]	Horizontal recommended	None	[illegible]
[illegible]	Horizontal recommended	None	1.0
[illegible]	Horizontal recommended	None	1.0
[illegible]	[illegible]	115 V, 60 Hz, 50 W	[illegible]
[illegible]	Any position	[illegible] 200 W	6.0
[illegible]	Any position	[illegible] 100 W	[illegible]
[illegible]	[illegible] recommended	[illegible] 20 W	[illegible]
[illegible]	[illegible] recommended	[illegible] 30 W	[illegible]
[illegible]	Any position	[illegible]	[illegible]

7.2 Types of flow-measuring instruments

Instruments used in the measurement of flow may be categorised into two main classes:

(i) QUANTITY METERS. In this class of instruments, total quantity which flows in a given time is measured and an average flow rate is obtained by dividing the total quantity by time.

(ii) FLOW METERS. In this class of instruments, actual flow rate is measured. Flow rate measurement devices frequently require accurate pressure and temperature measurements, in order to calculate the output of the instrument and the overall accuracy of the instrument which depends on the accuracy of pressure and temperature measurements. Quantity meters are used for calibration of the flow meters.

The classification can be further done as follows:

(i) QUANTITY METERS
- (a) Weight or volume tanks
- (b) Positive displacement or semi-positive displacement meters.

(ii) FLOW METERS
- (a) *Obstruction meters*
  - (i) Orifice
  - (ii) Nozzle
  - (iii) Venturi
  - (iv) Variable-area meters
- (b) *Velocity probes*
  - (i) Static pressure probes
  - (ii) Total pressure probes
- (c) *Special methods*
  - (i) Turbine type meters
  - (ii) Magnetic flow meters
  - (iii) Sonic flow meters
  - (iv) Hot wire/film anemometers
  - (v) Laser anemometers
  - (vi) Mass flow meters
  - (vii) Vortex shedding phenomenon
- (d) *Flow visualisation methods*
  - (i) Shadowgraphy
  - (ii) Schlieren photography
  - (iii) Interferometry.

This does not exhaust the list of flow-measuring systems but does attempt to include those systems which are of general interest. Special emphasis will be given to those instruments which find importance both in research and industry. The range, accuracy, operating conditions and other parameters of some of these instruments are shown in Table 7.1.

QUANTITY METERS

This type of meter gives an indication which is proportional to the total quantity that has flown in a given time. They are used for the flow measurement of both liquids and gases. A wide variety of these instruments are available but only positive displacement type instruments are discussed here. They all 'chop' the flow into 'pieces' of known size (known volume) and then count the number of 'pieces'. Fig. 7.1 shows

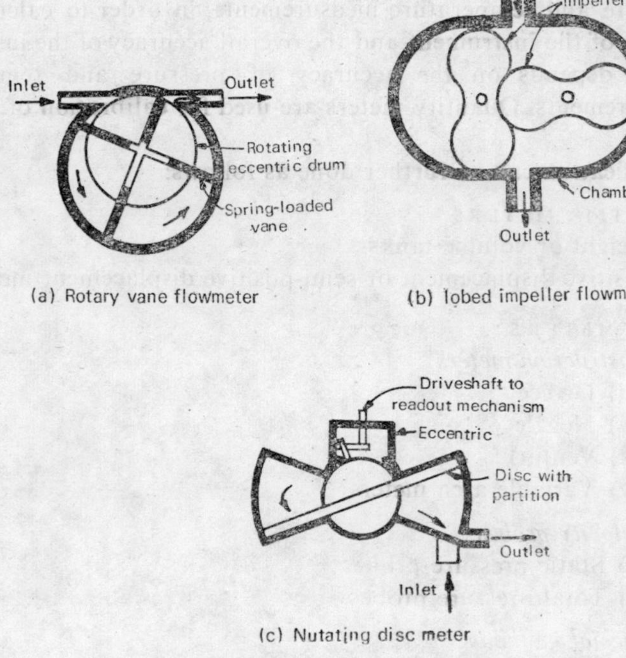

Fig. 7.1 Positive displacement meters

a number of positive displacement meters. These are often used for the measurement of volumetric flow of water, high viscosity liquid and liquids of varying viscosity. Their accuracy can be very good even at low end of the flow range. Since they are devices with moving parts, their accuracy may suffer with time due to wear. The accuracy can be enhanced by machining moving parts with fine clearances. Resistance to corrosive liquids can be increased by using special material, both for case and moving parts. Measurement of liquid with entrained vapour can be a very big problem, which is solved by inserting good vapour traps in the flow pipe. Most of these instruments are totalizers and do not attempt to measure instantaneous flow rates. The flow is transduced to rotary motion.

In short one can say that displacement meters are hydraulic or pneumatic motors whose cycles of motion are recorded by some form of counter. Energy is extracted from the flow to drive these meters, resulting in a pressure loss from inlet to exit of the instruments. But the energy required is extremely low; just enough to overcome friction in the system.

Obstruction Meters or Head Meters

By the law of conservation of mass, ideally when a fluid flows through a pipe with a restriction, the rate of flow increases due to the decrease in area. The physical relationship is such that the static pressure decreases as the rate (velocity) of flow increases. The pressure drop is an indication of the flow rate. Thus the obstruction transduces the velocity into a pressure change. Consider a one-dimensional flow in a pipe as shown in Fig. 7.2. The continuity equation for this situation demands that

$$\dot{m} = \rho_1 A_1 u_1 = \rho_2 A_2 u_2$$

where $\dot{m}$ is mass flow rate, ρ's the densities at planes 1 and 2 and u's the flow velocities. If the flow is adiabatic and frictionless and fluid is incompressible ($\rho_1 = \rho_2$), the Bernoulli equation, which governs the flow, may be written as

$$p_1 - p_2 = \frac{\rho}{2g_c}(u_2^2 - u_1^2)$$

Substituting for u_1 from the continuity equation,

$$p_1 - p_2 = \frac{\rho u_2^2}{2g_c}\sqrt{1 - \left(\frac{A_2}{A_1}\right)^2}$$

The volumetric flow rate is given by

$$Q = A_2 u_2 = \frac{A_2}{\sqrt{1 - \left(\frac{A_2}{A_1}\right)^2}}\sqrt{\frac{2g_c}{\rho}(p_1 - p_2)}$$

The volumetric flow rate is thus proportional to the square root of pressure drop and is a function of other known parameters. Therefore, a channel like the one shown in Fig. 7.2 can be used for flow measurement by measuring pressure differential.

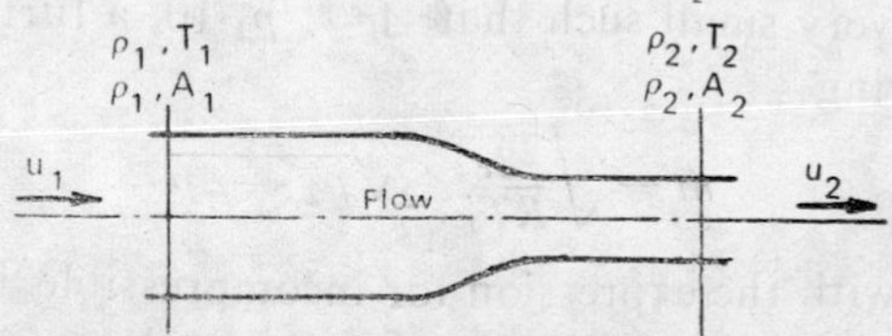

Fig. 7.2 General one-dimensional flow system

Discharge coefficient

In this derivation of the above equation, it is assumed that the channel is frictionless. However, no channel is frictionless and some losses are always present in the flow. The actual flow rate is then different from that calculated on the basis of the above equation. The actual flow rate is related to the ideal one through the following relation:

$$C = \frac{\dot{Q}_{\text{actual}}}{\dot{Q}_{\text{ideal}}}$$

where C is known as discharge coefficient. The discharge coefficient is not a constant and may depend strongly on Reynolds number ($< 15{,}000$) and the channel geometry.

Compressible fluid

The flow of a compressible fluid, say an ideal gas, obeys the following equation of state

$$p = \rho RT$$

where R is the gas constant and T is the absolute temperature. For reversible adiabatic flow, the steady flow energy equation is

$$c_p T_1 + \frac{u_1^2}{2g_c} = C_p T_2 + \frac{u_2^2}{2g_c}$$

where c_p is the specific heat at constant pressure and is constant for an ideal gas. Assuming a very small approach velocity ($u_2^2 > u_1^2$) and applying the continuity and state equations, yields the equation

$$\dot{m}^2 = 2g_c A_2^2 \frac{\gamma}{\gamma - 1} \frac{p_1^2}{RT_1} \left(\frac{\rho_2^2}{\rho_1^2} - \frac{T_2 \rho_2^2}{T_1 \rho_1^2} \right)$$

where $\gamma = c_p/c_v$ is adiabatic constant. This equation can be rewritten as

$$\dot{m}^2 = 2g_c A_2^2 \frac{\gamma}{\gamma - 1} \frac{p_1^2}{RT_1} \left[\left(\frac{p_2}{p_1} \right)^{2/\gamma} - \left(\frac{p_2}{p_1} \right)^{(\gamma+1)/\gamma} \right]$$

Compressible vs. incompressible fluid flow

The above equation can be simplified further and is given as follows:

$$\dot{m} = \sqrt{\frac{2g_c}{RT_1}} A_2 \left[p_2 \, \Delta p - \left(\frac{1.5}{\gamma} - 1 \right) \Delta p^2 + \ldots \right]^{1/2}$$

This equation holds good when $\Delta p = (p_1 - p_2) < p_1/4$. However, if the pressure drop is very small such that $\Delta p < p_1/10$, a further simplification may be made giving

$$\dot{m} = \sqrt{\frac{2g_c}{RT_1}} A_2 \sqrt{p_2 - \Delta p}$$

Comparing this with the expression for incompressible fluid flow, it may be concluded that for small values of pressure drop, Δp, compared with p_1, the flow of compressible fluid may be approximated by the flow of an incompressible fluid.

Types of Restriction Meters

The restriction provided in the flow passage for the purpose of flow metering is the primary element; three popular types exist and are called:

(i) Orifice plate,

(ii) Flow nozzle, and
(iii) Venturi tube.

These are shown in Fig. 7.3 along with the curves for pressure variation along the channel.

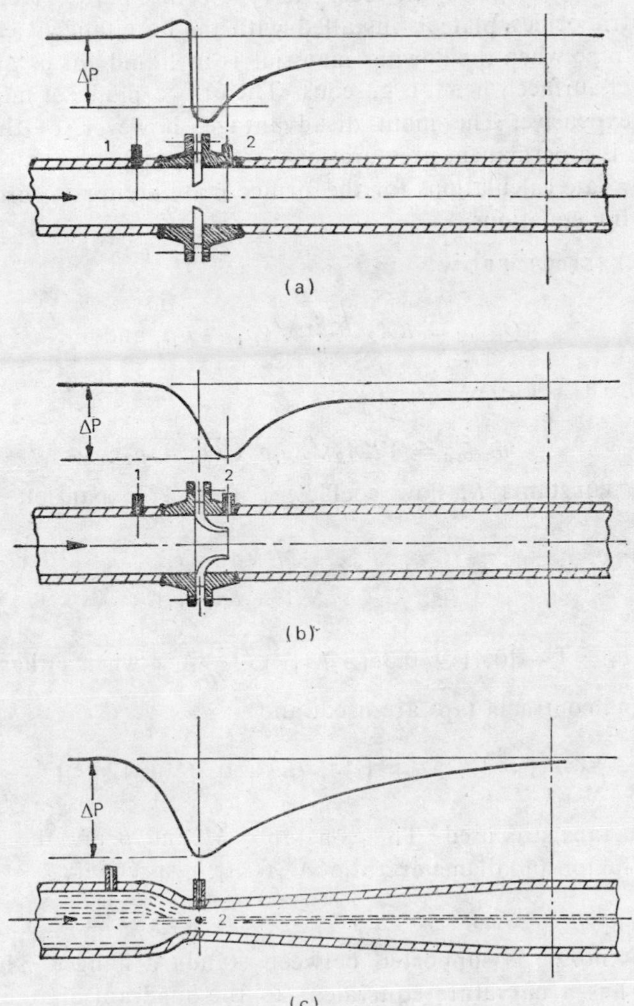

Fig. 7.3 (a) An orifice plate, (b) A flow nozzle, (c) A venturi

(i) *Orifice plate*

It is a thin, flat disc, with an orifice for the passage of fluid, and is inserted between flanges in the pipe. It can be readily rebored or replaced to accommodate flow capacity changes. Generally it is furnished so that the orifice is concentric; it is sometimes provided with an additional small hole for the passage of condensates and gases. When gas is metered, the hole is located at the bottom to allow the condensates to pass in order to prevent its build-up at the orifice. When the fluid is in liquid form, the extra hole is at the top to permit the gas to pass, thus avoiding the build-up of gas pockets. The presence of such a drain hole is often considered a disadvantage because it may introduce measurement errors. Sometimes, the orifice may be eccentric or segmental. When liquids contain high percentage of dissolved gases, the eccentric type orifice plate is recommended. The eccentric orifice plate is installed with the bore tangent to the upper surface of pipe when the flowing material is in liquid form, and tangent to the lower surface when it is gaseous. The orifice plates of moderate size are least expensive. The main disadvantage, however, of this kind of restriction is a permanent pressure loss, often upto 30%.

The flow rate calculations for the orifice plate are made on the basis of the following equations:

INCOMPRESSIBLE FLOW

$$\dot{Q}_{\text{actual}} = KA_2\sqrt{\frac{2g_c}{\rho}}\sqrt{(p_1 - p_2)}, \text{ and}$$

COMPRESSIBLE FLOW

$$\dot{m}_{\text{actual}} = YKA_2\sqrt{2g_c\rho_1}\sqrt{(p_1 - p_2)};$$

where the constants K, flow coefficient, and Y, expansion factor are defined as

$$K = CM = C\frac{1}{\sqrt{1 - (A_2/A_1)^2}}$$

and $Y = Y_1 = 1 - [0.41 + 0.35(A_2/A_1)^2]\dfrac{p_1 - p_2}{\gamma p_1}$, when either the flange taps or vena contracta taps are used, and

$$Y = Y_2 = 1 - [0.333 + 1.145\,(\beta^2 + 0.7\beta^5 + 12\beta^{13})]\,\frac{p_1 - p_2}{\gamma p_1}$$

when pipe taps are used. The constants M and β are the velocity of approach factor and diameter ratio, $\sqrt{A_2/A_1}$, respectively.

(ii) *Flow nozzle*

The flow nozzle is supported between standard flanges. The rounded approach has a curvature equivalent to the quadrant of an ellipse. The curved surface guards the nozzle from corrosive/erosive effects due to the suspensions in the gas, thus contributing to its long life. It allows measurement of flow rates which are about 60 to 65% higher than the maximum flow rate for which an orifice plate can be used; the flow nozzle will find its main application where under high operating pressures great capacities must be measured through lines which are reduced to minimum size for some reason. Another advantage of using nozzle is that it requires smaller straight piping before and after the primary element compared to that of orifice. The pressure loss between that of orifice and venturi, mainly is due to the absence of recovery cone.

The flow rate calculations are made on the basis of following equations:

INCOMPRESSIBLE FLUIDS

$$\dot{Q}_{\text{actual}} = KA_2\sqrt{\frac{2g_c}{\rho}}\sqrt{(p_1 - p_2)}$$

COMPRESSIBLE FLUIDS

$$\dot{m}_{\text{actual}} = YKA_2\sqrt{2g_c\rho_1}\sqrt{(p_1 - p_2)}$$

where the expansion factor Y is given by

$$Y = \left[\left(\frac{p_2}{p_1}\right)^{2/\gamma} \frac{\gamma}{\gamma - 1} \frac{1 - (p_2/p_1)^{(\gamma-1)/\gamma}}{1 - (p_2/p_1)} \frac{1 - (A_2/A_1)^2}{1 - (A_2/A_1)^2(p_2/p_1)^{2/\gamma}}\right]^{1/2}$$

(iii) *Venturi tube*

The venturi tube offers the best accuracy, least pressure loss ($\simeq$ 13%) and best resistance to abrasion and wear from dirty fluids. It is, however, expensive and occupies substantial space. Due to its excellent pressure recovery characteristics, it is recommended where measuring conditions require extremely low pressure loss. Because of its streamlined approach and exit, use of venturi tubes is often considered when the flow of liquid with solids in suspension must be measured. In a venturi tube, there is an appreciable distance between the pressure taps; the fluid friction along a side wall of pipe which varies as the fifth power of the diameter can have significant effect on the pressure differential. Therefore, when suspensions are carried in the fluid, the fluid friction depends on the concentration of suspension and hence pressure differential will exhibit sensitivity to the concentration of suspensions.

The flow rate calculations for venturi tube are made on the basis of following equations:

INCOMPRESSIBLE FLUIDS

$$\dot{Q}_{\text{actual}} = CM\, A_2\sqrt{\frac{2g_c}{\rho}}\sqrt{(p_1 - p_2)}, \text{ and}$$

COMPRESSIBLE FLUIDS

$$\dot{m}_{\text{actual}} = Y\, CM\, A_2\sqrt{2g_c\rho_1}\sqrt{(p_1 - p_2)}$$

PRACTICAL CONSIDERATIONS

The construction of obstruction meters has been standardised by organisations like ASME, DIN and ISI. The recommendations are briefly discussed here:

Orifice

The recommended installations for concentric, thin plate, square edged orifice are shown in Fig. 7.4 (a). Note that three standard pressure-tap locations are used:

(i) FLANGE TAP. Installed in the flanges and most universally used.

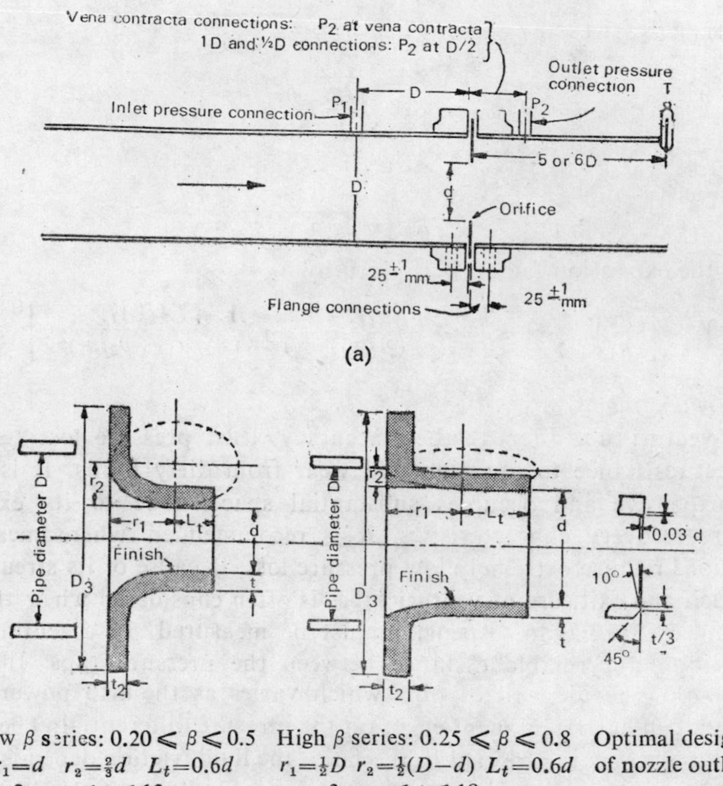

(a)

Low β series: $0.20 \leqslant \beta \leqslant 0.5$ High β series: $0.25 \leqslant \beta \leqslant 0.8$ Optimal designs of nozzle outlet

$r_1 = d$ $r_2 = \frac{2}{3}d$ $L_t = 0.6d$ $r_1 = \frac{1}{2}D$ $r_2 = \frac{1}{2}(D-d)$ $L_t = 0.6d$

$3\text{ mm} \leqslant t \leqslant 13\text{ mm}$ $3\text{ mm} \leqslant t \leqslant 18\text{ mm}$

$3\text{ mm} \leqslant t_2 \leqslant 0.15D$ $3\text{ mm} \leqslant t_2 \leqslant 0.15D$

(b)

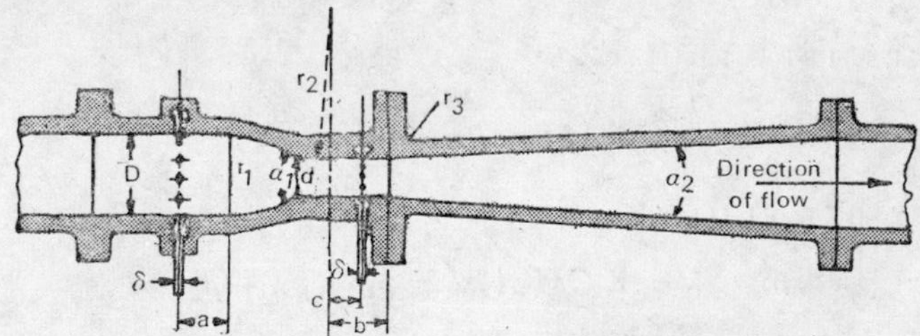

D = Pipe diameter inlet and outlet d = Throat diameter as required

$a = 0.25D$ to $0.75D$ for $100\text{ mm} \leqslant D \leqslant 150\text{ mm}$,
$0.25D$ to $0.50D$ for $150\text{ mm} < D \leqslant 800\text{ mm}$

$b = d$ $c = d/2$

δ = 6 mm to 12 mm according to D. Annular pressure chamber with at least four piezometer vents

$r_1 = 0$ to $0.25D$ $r_2 = 0$ to $0.25d$ $r_3 = 0$ to $0.25d$

$\alpha_1 = 21° \pm 1°$ $\alpha_2 = 7°$ to $15°$

(c)

Fig. 7.4. Recommended installations: (a) Concentric, thin plate, square edged orifice; (b) Long radius flow nozzles; (c) Venturi tubes.

(ii) PIPE TAPS. The inlet pressure tap is located one pipe diameter upstream and the outlet tap is located half diameter downstream of the orifice.

(iii) VENA CONTRACTA TAPS. The inlet pressure tap is located one pipe diameter upstream, and the outlet pressure tap is located at vena contracta. Such taps are employed mostly in large size pipes where the use of a flange union is impractical.

The orifice discharge coefficient is sensitive to the condition of the upstream edge of the hole. The discharge coefficient is the same for liquids or gases as long as the Reynolds number is same. Fig. 7.5 shows the dependence of flow coefficients, MC, on Reynolds number for various values of β the diameter ratio, for flat-plate orifice.

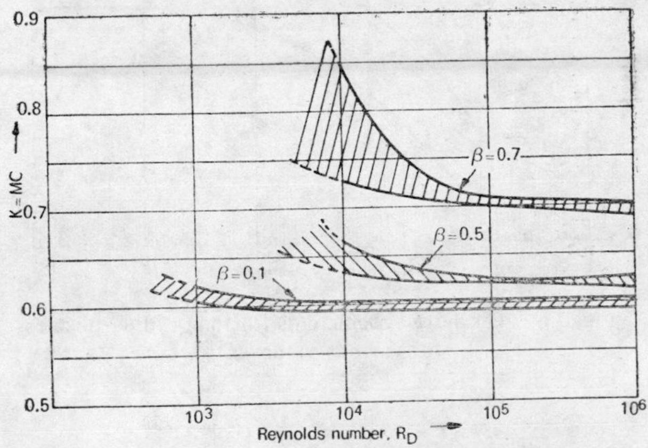

Fig. 7.5. Dependence of flow coefficient MC on Reynolds number R_D for various values of β for flat-plate orifices

Flow nozzles

The recommended proportions for the flow nozzles are given in Fig. 7.5 (b). The approach curve must be proportional to prevent separation between flow and the wall, and parallel section is used to ensure that the flow fills the throat. Usually pipe taps are used. Fig. 7.6 illustrates the dependence of discharge coefficient on Reynolds number for various values of β for a long radius nozzle.

Venturi tube

The recommended proportions of a standard venturi tube are shown in Fig. 7.4 (c). Note that the pressure taps are connected to manifolds which surround the upstream and throat portions of the tube. These manifolds receive a sampling of pressure all around the sections so that a good average value is obtained. The discharge coefficient as a function of Reynolds number for a venturi are shown in Fig. 7.7 with the tolerance limits indicated by the dotted lines.

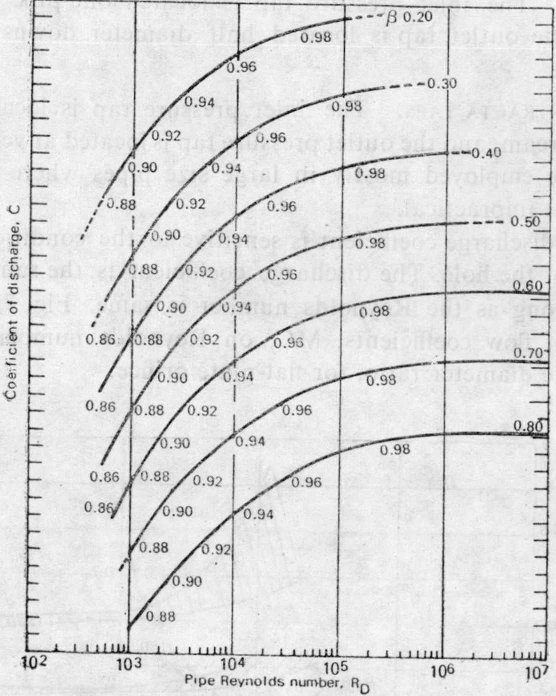

Fig. 7.6. Discharge coefficients for long radius nozzles for various values of β

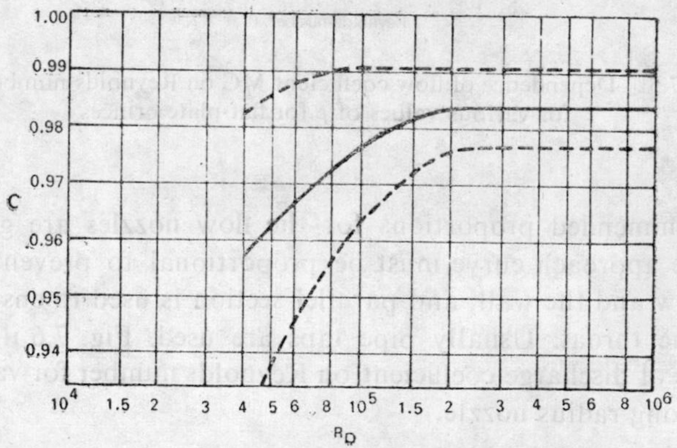

Fig. 7.7 Discharge coefficient for venturi tubes as a function of Reynolds number

When the compressible fluids are to be metered, the expansion coefficient should be calculated from the relations given earlier and the flow rate calculated. The variation of expansion coefficient, Y, for orifices, nozzles and venturi tubes as a function of pressure differential is given in Fig. 7.8. Note the linear dependence of Y on $(p_1 - p_2)$ for whole range of β

values. The measurement of flow rates by the obstruction meters reduces to measurement of pressure differential.

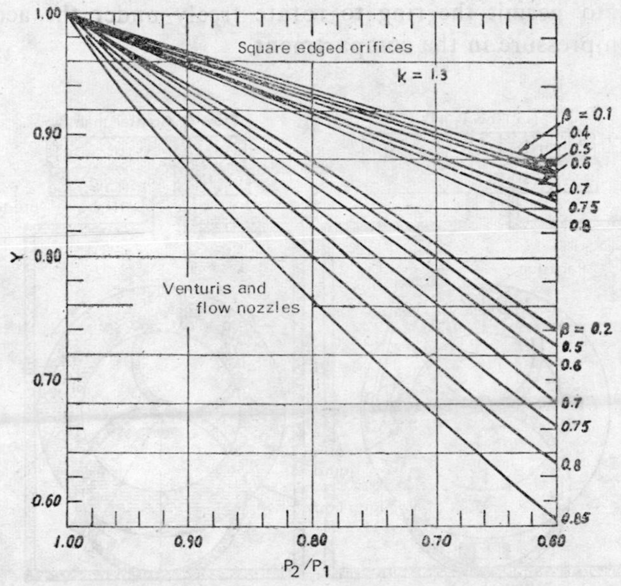

Fig. 7.8 Variation of expansion coefficient, Y, with pressure ratio

7.3 Measurement of pressure differential

The determination of flow rate using obstruction meters involves the measurement of pressure differential. Many meters are available, for example, mercury float flow meter, Ledoux-bell meter, ring-balance meter, aneroid meter, force-balance type meter, inductance bridge flow meter, etc., wherein pressures from the upstream and downstream sides of the primary element are fed in simultaneously. Some of the meters extract the square roots, like Ledoux-meter, so that the output is linear, in others square root chart paper or scale, like in ring-balance meter, is used. One such meter, ring-balance meter, is discussed in detail here.

Ring-balance Meter

The ring-balance meter is a radial torque meter which uses a hollow ring-body to convert the pressure differential generated by a differential medium, or by a difference in static pressure, into a rotation which is transmitted to the recorder or indicator.

Figure 7.9 shows the ring assembly in a typical flow-measuring system. Fig. 7.9 (a) indicates the position of ring-body at pressure equalisation or zero flow condition, while Fig. 7.9 (b) indicates the position when pressure differential exists.

The ring assembly is mounted on a knife-edge bearing, which permits the rotation about the axis of the ring. The ring is divided into two pressure compartments by the baffle at the top, and by the sealing liquid which fills the lower part of the ring. The two ring compartments thus formed are connected to the differential pressure pipes by means of flexible tubing to permit the ring to rotate freely under the action of the difference in pressure in the compartment.

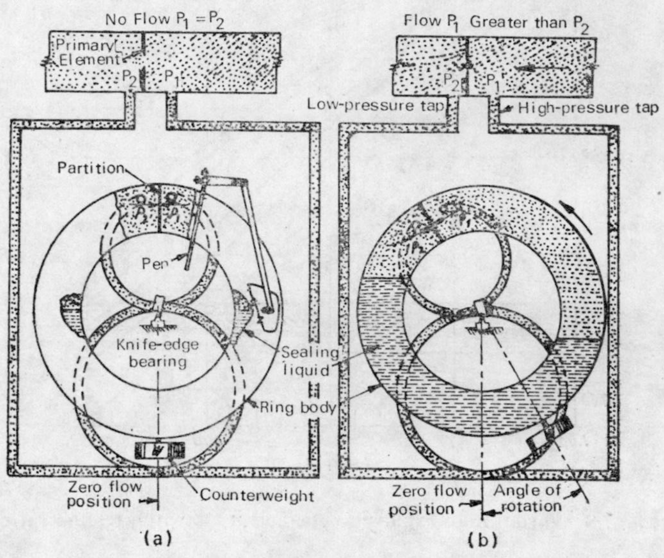

Fig. 7.9 Ring-balance meter body

The ring rotation is transmitted to the recording pen or indicating pointer by a linkage. When the meter is used to register flow rate, a square root chart or scale is employed. In some cases the ring rotation is transmitted to the pen through a cam and the cam follower. The cam follower imparts a movement to the pen which is directly proportional to the flow rate.

Ring torque is a function of differential pressure acting on the baffle. The torque is resisted by an external calibration weight, rigidly attached to the bottom of ring body. The meter is in equilibrium when the ring torque is balanced by this counter weight.

Mercury or other sealing fluid exerts no force tending to rotate the ring, but acts only as a seal for the differential pressure in the two compartments. This is true because in the circular body all hydrostatic forces resulting from the deflection of the sealing liquid are directed normally to the containing circle, and therefore pass radially through the exact centre of rotation without producing meter torque.

The ring-balance may contain two *S*-shaped, self-compensating tubes for relatively high static pressure or parallel tubes for lower pressure applications.

Variable-area Meters

In obstruction meters, restriction in the channel is of fixed size and the pressure differential across it changes with the flow rate. The flow rate is proportional to the square-root of pressure differential. This is often a disadvantage because for the measurement of wide ranges of flow rates, pressure measuring system of very wide range is required. In other words, if the range is accommodated, the sensitivity is not uniform over the whole range. On the other hand, in variable-area meters, the size of restriction is adjusted by an amount necessary to keep the pressure differential constant. The amount of adjustment required is a function of flow rate. There are two basic types of area meters—piston type meter and rotameter.

Piston type meter

It is designed specifically for metering viscous liquids, such as hot tar, black liquor, etc., which are difficult to measure in any other way. It is

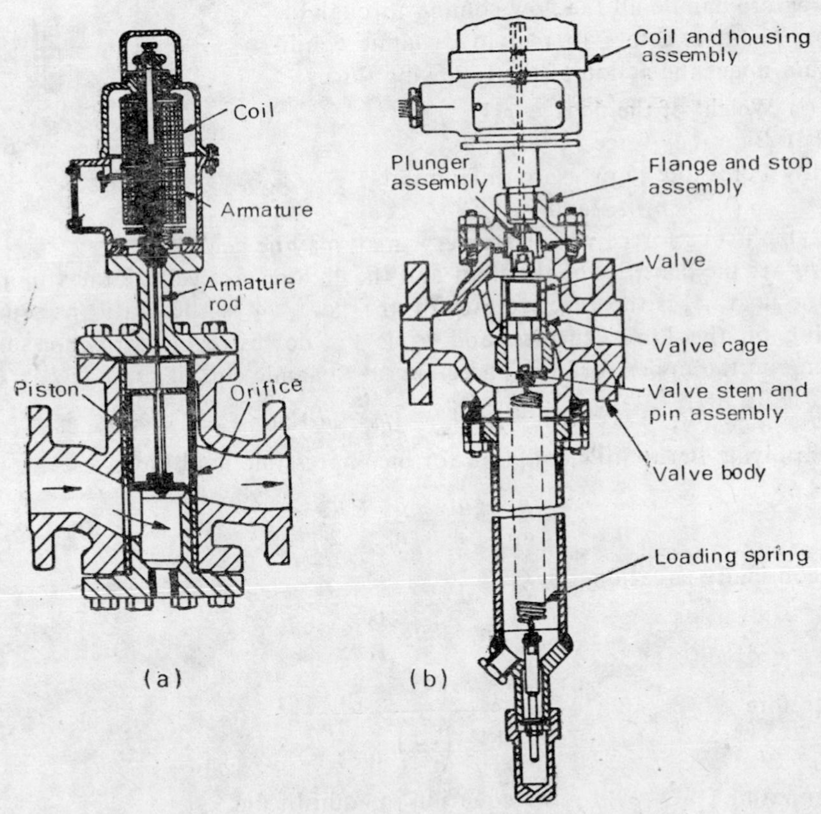

Fig. 7.10 (a) Piston type weight loaded area flow meter with electric transmission; (b) Piston type spring loaded area flow meter with electric transmission

installed directly in the pipeline and is usually equipped with an electric transmitter. If the rate of flow increases, the differential pressure across the metering plug tends to increase. This raises the plug and increases the port area in proportion to the rate of flow. The converse is true when the flow decreases. This is shown in Fig. 7.10.

Rotameter

The rotameter consists of a tapered tube, mounted vertically with the smaller end of the tube facing down, in which a metering float (bob) is located (Fig. 7.11). The fluid flows through the tube from bottom to top. When no fluid is flowing, the float rests at the bottom of tapered tube, and its maximum diameter is usually so selected that it blocks the small end of the tube almost completely. When flow commences, float rises till the annular passage between the inner wall of the tapered tube and periphery of the float is large enough to handle all the flow coming through the pipe. The float comes to rest in dynamic equilibrium under the action of the following forces:

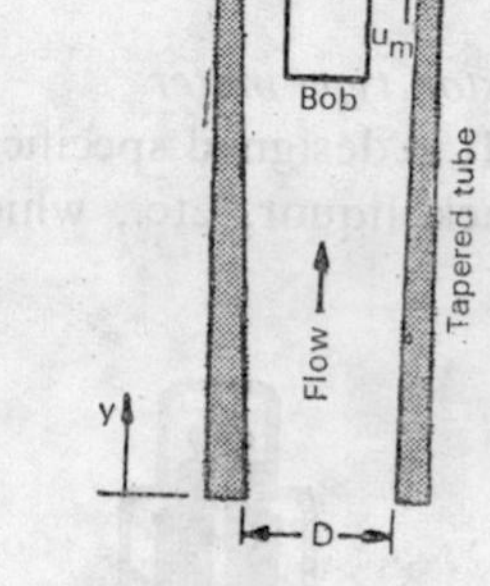

Fig. 7.11 Schematic of rotameter

(i) Weight of the float $= \rho_b Vg$,
(ii) Buoyancy force $= \rho_f Vg$,
(iii) Force due to pressure differential $= (p_1 - p_2)A_e$; and
(iv) Viscous forces which are very small may be neglected.

ρ_b, ρ_f are the densities of the float and the fluid respectively, V the volume of the float, A_e is the effective area of the float, p_1 is the static pressure acting on the lower surface and p_2 is the downstream static pressure acting on the upper surface. Therefore, at dynamic equilibrium

$$(p_1 - p_2)A_e = (\rho_b - \rho_f)Vg$$

By applying Bernoulli's equation for incompressible fluid flow,

$$\frac{p_i}{\rho_f} + \frac{u_1^2}{2} = \frac{p_2}{\rho_f} + \frac{u_2^2}{2}$$

By continuity equation,

$$u_1 = u_2 \frac{A_2}{A_1}$$

Therefore

$$\frac{u_2^2}{2} = \frac{1}{\left[1 - \left(\frac{A_2}{A_1}\right)^2\right]} \cdot \frac{p_1 - p_2}{\rho_f}$$

Eliminating $(p_1 - p_2)$ by the equation of equilibrium

$$u_2 = \frac{1}{\sqrt{1 - \left(\frac{A_2}{A_1}\right)^2}} \cdot \sqrt{\frac{2Vg}{A_e} \cdot \frac{(\rho_f - \rho_b)}{\rho_f}}$$

Hence, for actual rate of flow

$$\dot{Q} = \frac{C_d A_2}{\sqrt{1 - \left(\frac{A_2}{A_1}\right)^2}} \cdot \sqrt{\frac{2Vg}{A_e}} \cdot \sqrt{\frac{\rho_f - \rho_b}{\rho_f}}$$

where $A_2 = \frac{\pi}{4}[(D + \alpha y)^2 - d^2] \simeq \frac{\pi}{4} D\alpha y$

D is the diameter of the tube at the inlet, d is the maximum diameter of the float, usually $D \simeq d$, α is the constant indicating the taper of the tube and y is the vertical distance of float from the entrance. The effect of drag force is included in the coefficient C_d, called drag coefficient. Therefore,

$$\dot{Q} = C'y\sqrt{\left(\frac{\rho_b}{\rho_f} - 1\right)}$$

where C' is a constant.

Thus every float position, y, corresponds to one particular flow rate and no other. It is necessary to provide a reading or linear calibration scale on the outer side of the tube, and flow can be determined by direct observation of the metering float.

The equation of mass flow rate is given by

$$\dot{m} = \dot{Q}\rho_f = C'y\sqrt{\rho_f(\rho_b - \rho_f)}$$

It is frequently advantageous to have a rotameter which gives an indication independent of change in fluid density, due to changes in temperature. By optimizing

$$\rho_b = 2\rho_f$$

and the mass flow rate is given by

$$\dot{m} = C'y\,\frac{\rho_b}{2}$$

Thus by constructing the float from a material satisfying the above condition, the meter may be used to compensate for density changes in the fluid due to temperature changes. The error in metering $\dot{m}$ under this condition is less than 0.2% for a fluid density variation of 5% from the above condition.

7.4 Velocity probes

When the description of a flow field is desired, both magnitude and direction of flow-velocity vector at various points in the field should be known. This is achieved by variety of pressure probes; measurement of flow velocity vector is carried out over a finite area due to the finite dimensions of probe, instead of on a 'point'. Thus the direct measurement results in the values of average flow conditions.

The choice of a particular probe rests on many factors such as the type of information required, size flow conditions, etc. Basically, pressure probes measure either of the two different pressures or some combination thereof. These are static p_s and total p_t pressures, such that

$$p_t = p_s + p_v$$

where p_v is the velocity pressure.

Substituting for velocity pressure and rearranging an expression for the flow velocity is given as

$$v = \sqrt{\frac{2g(p_t - p_s)}{\rho}}$$

where ρ is the density of fluid. Therefore, the velocity may be determined simply by measuring the differences between total and static pressures.

The above equation holds good for incompressible fluid. For compressible flow, the above relation is suitably modified. Assume an ideal gas (compressible fluid) undergoing an isoentropic process, then $p/\rho^\gamma = \text{const.}$ Thus

$$\left(\frac{p_t}{\rho_t} - \frac{p_s}{\rho_s}\right)_{\text{comp.}} = \left(\frac{\gamma - 1}{\gamma}\right)\frac{v^2}{2g}$$

where subscript 'comp.' means compressible. This relation can be expanded in terms of Mach number M. Thus

$$(p_t - p_s)_{\text{comp.}} = \frac{\rho v^2}{2g}\left(1 + \frac{M^2}{4} + (\gamma - 2)\frac{M^4}{24} + \ldots\right)$$

where $M \to 0$; the above relation reduces to that of incompressible fluid.

The effect of compressibility can usually be taken into account through the use of a correction factor c', thus

$$v = (1 - c')\sqrt{\frac{2(p_t - p_s)g}{\rho}}$$

Fig. 7.12 shows the variation of c' with velocity based on air at atmospheric pressure.

When the velocity is used to measure flow rate, a weighted velocity is

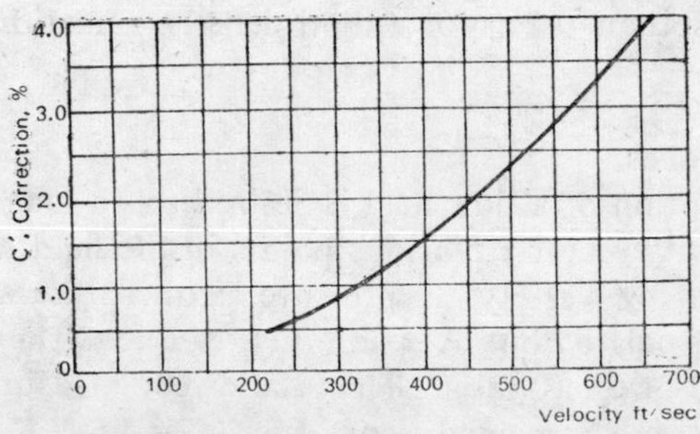

Fig. 7.12 Velocity correction for fluid compressibility

to be obtained by measuring it at various points in the channel, or a multiplication factor may be determined by calibration, for a given Reynolds number.

The static pressure can be sensed in the following ways:

(i) WALL TAPS. Small holes can be drilled in the surface of the flow boundary in such a way that streamlines of the flow remain relatively undisturbed. The accuracy of measuring static pressure with wall taps is determined by the size and shape of the hole. It has been shown by Shaw that for a smooth pipe with incompressible turbulent flow and a static pressure hole diameter of 1/10 of the pipe diameter, the static pressure error reaches about 1% of the mean dynamic pressure at a pipe Reynolds number of 2×10^6.

(ii) STATIC TUBES. The accuracy in static pressure measurement using static tubes depends on the position of the sensing holes with respect to the nose of the tube and main supporting stem. Streamlines next to the nose of the tube must be longer than those in the undisturbed flow, indicating an increase in the velocity. Acceleration effects thus caused by the nose tend to lower the tap pressure; while the stagnation effects caused by the stem tend to raise the tap pressure. The static tube characteristics are shown in Fig. 7.13.

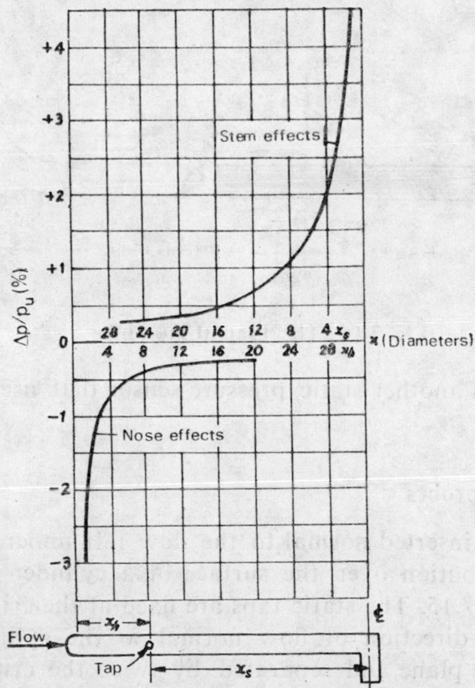

Fig. 7.13 Static tube characteristics

In a properly compensated tube, the acceleration and stagnation effects will just balance each other at the plane of the pressure holes.

This is the principle of Prandtl-Pitot tube. Fig. 7.14 shows the design of this tube which utilises eight square edged pressure holes (1 mm in diameter) placed 45° apart in a plane located 8 tube diameters back from the nose and 16 tube diameters forward from the probe stem.

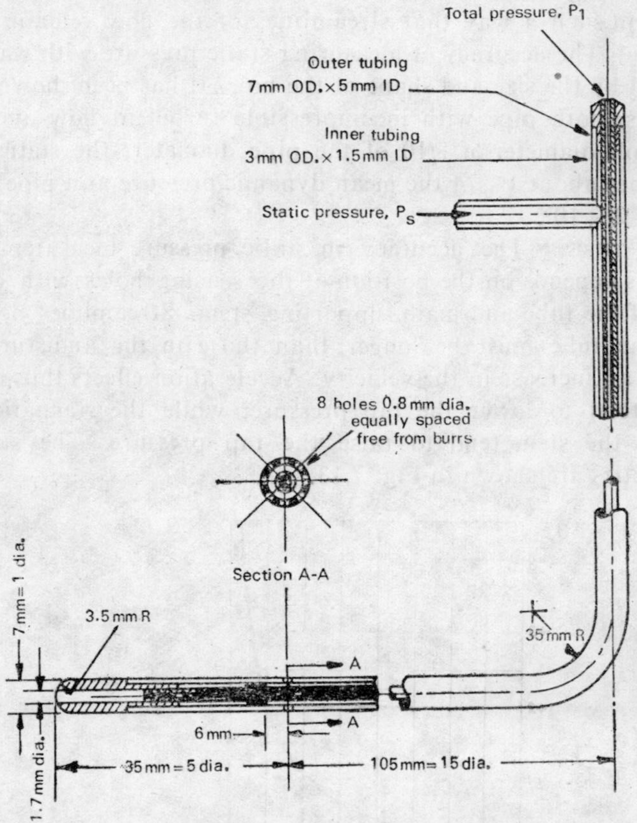

Fig. 7.14 The Prandtl-Pitot tube

The disc probe is another static pressure sensor that uses the compensation principle.

7.5 Aerodynamic probes

Cylindrical probes inserted normal to the flow fall under this category. The pressure distribution over the surface of a cylinder is well known and shown in Fig. 7.15. The static taps are fixed at the critical angle. In order to find the direction of flow normal to the cylinder, two taps located in the same plane and separated by twice the critical angle are used. This is accomplished by rotating the cylinder until the two pressure taps, connected across a manometer, sense identical pressures. The wedge probe is also used for this purpose. Figure 7.16 illustrates the comparison of performance of cylindrical and wedge probes. It is obvious that the wedge probe has a less rapid change in tap pressure in the region of pressure taps than does the cylindrical probe. Unfortunately, the fragile

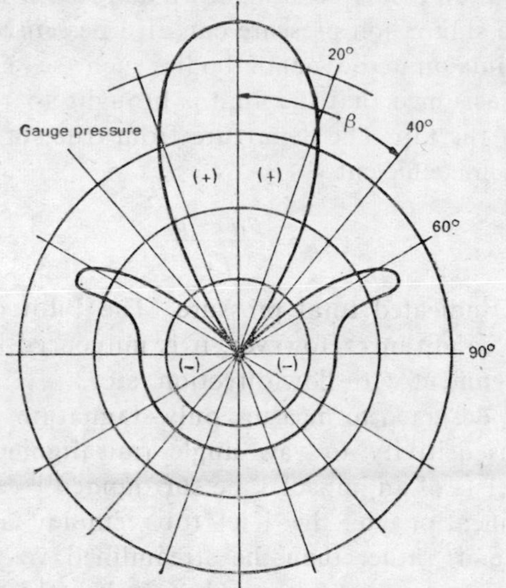

Fig. 7.15 Pressure distribution on surface of cylinder inserted normal to flow

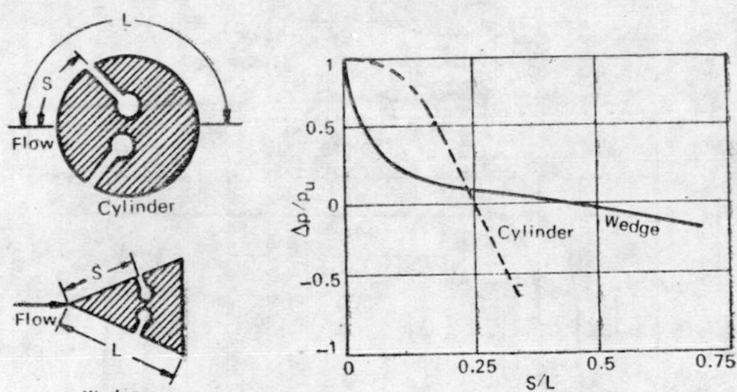

Fig. 7.16 Comparison of performance of cylinder and wedge-shaped aerodynamic probes

apex of the wedge makes it a less robust instrument and favours the use of cylindrical probe in many applications. The static pressure probes are used for measuring static pressures in obstruction meters. They are required in velocity determinations to establish thermodynamic state points and are also useful in obtaining indications of flow direction.

7.6 Total pressure probes

The measurement of total or stagnation principle is usually somewhat easier than the measurement of static pressure. By definition the total pressure can be sensed by stagnating the flow isoentropically. The stagnation pressure can be measured adequately with the classical Pitot tube. More often, a Pitot tube is provided with the static openings; one such using a compensation principle and known as *Prandtl-Pitot tube* is shown in Fig. 7.14. The stagnation pressure can also be sensed by holes located at stagnation points on aerodynamic bodies such as spheres and cylinders. In all these it is assumed that the fluid is brought to rest isoentropically in the vicinity of the tap. The departure from true stagnation pressure is indicated by Mach-coefficient

$$c_p = \frac{p_{tI} - p_s}{p_v}$$

where p_{tI} is the indicated total pressure. The Pitot coefficient is unity under usual flow conditions. However, it is influenced by viscosity, probe geometry, misalignment with flow direction, etc.

The Kiel tube, designed to measure only stagnation pressure possesses a remarkable insensitivity to yaw angle (misalignment) as shown in Fig. 7.17. It consists of an impact tube surrounded by what is essentially a venturi. Modifications of the Kiel tube employ a cylindrical duct, bevelled at each end, rather than the streamlined venturi. This appears to have a little effect on performance but makes the construction quite cheap.

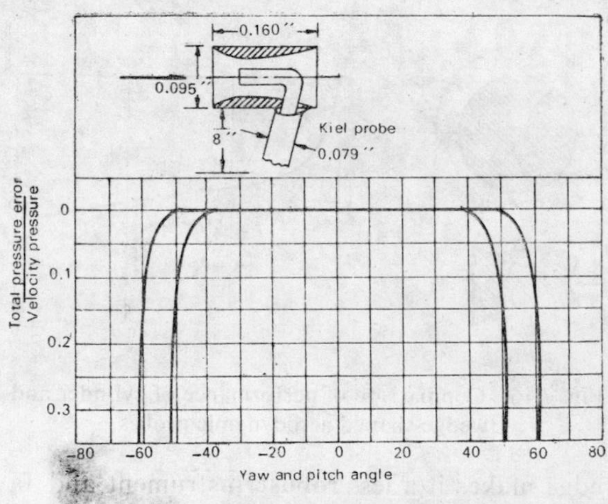

Fig. 7.17 The Kiel tube

7.7 Special methods

(i) Turbine Type Flow Meter

As the fluid flows through this meter, it causes a rotation of a small turbine wheel. In the turbine body is a permanent magnet which rotates with the wheel. A reluctance pick up is mounted at the body of the meter which picks up the pulse for each rotation of the wheel. Since the volumetric flow is proportional to the number of revolutions of the wheel, the total number of pulses may be taken as an indication of total flow. The pulse rate is proportional to flow rate. A flow coefficient K of the turbine meter may be defined:

$$\dot{Q} = f/K$$

where f is the pulse frequency and $\dot{Q}$ is the volume flow rate. The flow coefficient is dependent on flow rate and viscosity.

The commercial turbine meters can handle flow rate in a wide range commencing from 0.5 lpm to 150,00 lpm for liquid and 3 lpm to 500,000 lpm for air. The flow is indicated within an accuracy of $\pm 0.5\%$ over a wide range of flow rates. The transient response of the meter is quite good.

(ii) Electromagnetic Flow Meter

Electromagnetic flow meters are based on the Faraday principle of induction. If a conductor of length l moves with a transverse velocity v across a magnetic field of intensity B, an emf e is induced; the emf e is expressed as

$$e = Blv \times 10^{-8} \text{ volts}$$

where e = emf in volts

B = magnetic field in gauss

l = length in cm

v = velocity in cm/s

In practice the length of the conductor is equal to the diameter of the pipe. For less conducting fluid, like water, a non-conducting pipe-line is used and the electrodes are flushed with it so that they are in contact with the fluid. The pipe is of non-magnetic material as to allow the magnetic field to penetrate. The flow is usually assumed to be uniform; however, the result holds good for all symmetrical flows but then the average flow velocity is indicated.

For highly conducting fluids, like mercury, the metallic pipe (stainless steel is used as the pipe metal) is not very effective as a 'short circuit' for the voltage induced in the fluid flow.

Both d.c. and a.c. magnetic fields can be used—the d.c. magnetic field is preferred for metallic flow as the output is high and no amplification is required. While for liquids which can be polarised, an a.c. field is used and further amplification is possible which is usually required for less conducting fluids.

The main limitation of electromagnetic flow meters is the conductivity of the fluid—it must be sufficiently high so that the external circuitry is not an excessive load. However, it possesses the advantages of complete absence of any obstruction in the pipe; ability to measure reverse flows; insensitivity to viscosity, density and any other flow disturbance so long the flow is symmetrical; wider linear range and good transient response.

(iii) Sonic Flow Meter

This flow meter is based on the principle of addition of velocities—the one of ultrasonic is vectorially added to that of flow field and the resultant velocity is measured. In this flow meter, pressure waves of frequency around 10 MHz sent from an ultrasonic source either a piezoelectric or magnetostrictive transducer located externally on the pipe and picked up by the receiver on the opposite side of the transducer. In fact one measures the time delay in this method. In another variant of this method, the beat frequency is measured. The signal travelling through the flowing medium suffers a frequency change, which is proportional to the flow velocity. The frequency of this signal is compared with that of the reference signal.

Another approach senses the deflection of the ultrasonic beam propagated transversely to the flow. With no flow, the two receivers pick up the signals of the identical strengths. With flow, the signal strength at one receiver will increase and on the other it will decrease.

The ultrasonic flow meters have the advantages of not using any flow obstruction; insensitivity to viscosity, temperature and density variations. They are suited for the flow measurement of liquids.

(iv) Hot Wire Anemometer

Hot wire anemometer is a device which is most often used in research applications to study varying flow conditions. This is used in two basic modes—constant temperature mode, and constant current mode.

A thin wire carrying a current is exposed to the flow field. The wire attains a temperature when Joule's heating i^2R (i is the current flowing through the wire of resistance R) just balances the convective heat loss from the wire. The convective film coefficient is a function of flow velocity and hence as the flow velocity changes, temperature of wire changes. In *constant temperature mode*, the temperature of the wire is kept constant by varying the current flowing through this and its magnitude is taken as a measure of flow velocity. While for *constant current mode*, the Joule heat i^2R is essentially kept constant and when the flow velocity changes, the temperature of the wire adjusts itself so that the equilibrium is reached. The temperature of wire is then taken as the measure of flow velocity. Temperature is measured in terms of the resistance change of the wire using conventional bridge circuitry.

A common hot wire anemometer consists of a thin tungsten wire of 8 μm diameter and 1 mm long, having a resistance of approximately 1 ohm and supported on a non-conducting structure. The probe is thus effectively a point size. The main disadvantages of the hot wire anemometer lie in its limited strength, calibration changes caused by the accumulation of impurities on the wire, vibration of wire resulting in quick damage and flutter effects. Unless the flow field is clean, its use is not recommended. It has, therefore, been used mainly in gases.

In liquids, a variation of it, often called thin film anemometer is used.

A thin film of platinum is coated on a pyrex glass wedge and the connections are taken through heavy silver plates. It is very robust, easy to clean and can be used at higher temperatures.

In equilibrium, the following equation holds good

$$i^2R_w = K_c hA(T_w - T_f)$$

where K_c = conversion factor from thermal to electrical quantities,

T_w = wire temperature,

T_f = Temperature of flow field,

h = film coefficient of heat transfer,

A = heat transfer area.

The heat transfer film coefficient h is mainly a function of flow velocity for a given flow density for a range of velocities; this function has a general form

$$h = c_0 + c_1\sqrt{u}$$

where u is the flow velocity. Therefore

$$i^2R_w = c_2 + c_3\sqrt{u}$$

where $c_2 = K_cA(T_w - T_f)c_0$

and $c_3 = K_cA(T_w - T_f)c_1$

A relation between i^2 and $\sqrt{u}$ must essentially be a straight line.

Constant temperature mode

For measurement of average or steady velocities, the constant temperature mode is often used. Fig. 7.18 shows a circuit diagram. For an accurate work, a given hot wire probe is to be calibrated in a fluid in which it is to be used later. That is, it is exposed to known velocities (measured accurately by some other means) and its output recorded over a range of velocities.

In the circuit of Fig. 7.18, the current through R_w remains essentially constant even when R_w changes because R_I is of the order of 2000 ohms while R_1, R_2, R_3 and R_w are of the order of 1–20 ohms each. In calibration the hot wire is exposed to some known value of flow velocity, say u_1 and the current i through it, is so adjusted by varying R_I as to give adequate sensitivity without any danger of its burning. Resistance R_w of the wire will come to a definite value depending on the flow velocity. Now the resistance R_3 is adjusted to balance the bridge. This adjustment is essentially a measure of wire temperature which is fixed for all velocities. The first point for calibration is thus i_1^2, $\sqrt{\bar{u}_1}$. Now the value of u is changed causing wire temperature and R_w to change and unbalancing the bridge. Then R_w, and thereby the wire temperature are restored to their original values by adjusting i by means of R_1 to a new value until the bridge balance is obtained. R_3 is never changed during whole

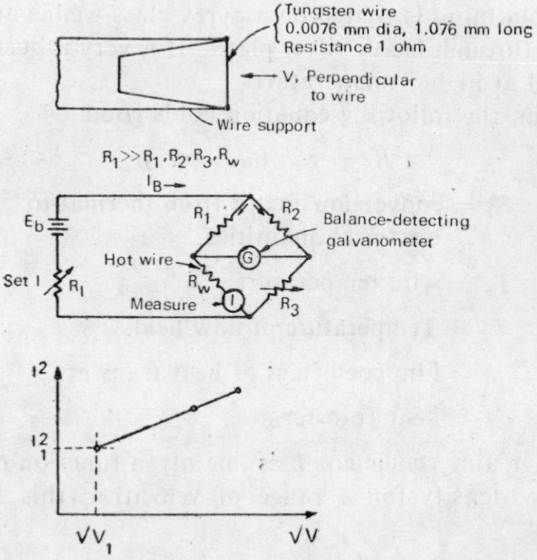

Fig. 7.18 Hot wire anemometer (Constant temperature model)

calibration procedure. The new current and the velocity are plotted as i^2 and $\sqrt{u}$ on the calibration curve. The procedure is repeated for as many velocities as desired. Once calibrated, the probe can be used to measure unknown velocities by adjusting R_I until the bridge balance is achieved, reading i and obtaining the corresponding u from the calibration curve. This assumes that the measured fluid is at the same temperature and pressure as the calibration fluid.

Constant current mode

Although the measurement of steady velocities is of some practical importance, perhaps the main application of hot wire anemometer is to study the fluctuating velocity flows. Both constant current and constant temperature modes can be used for this purpose. However, the constant current mode of operation is first discussed; the schematic is shown in Fig. 7.19. A fluctuating component of velocity over the average velocity $u(t) = u_0 + v(t)$ results in fluctuations in the resistance of the wire, consequently the voltage output is time-varying. The transient response of the system is first order and given by

$$\frac{e}{v}(D) = \frac{K}{\tau D + 1}$$

where e is the output voltage K is the static sensitivity and τ is the time constant. Time constant τ cannot be reduced beyond 0.001 s in actual practice which will limit the flat response to less than 160 Hz. This is quite inadequate for the study of turbulent flow where frequencies of the order of 50,000 Hz or higher are of interest. This limitation can be over-

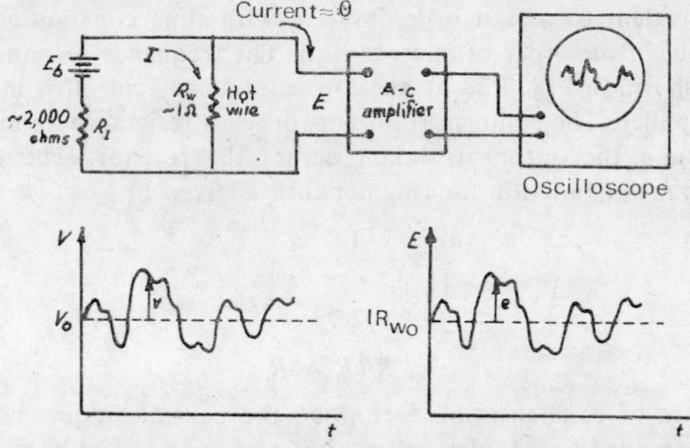

Fig. 7.19 Hot wire anemometer (Constant current mode)

come by electrical dynamic compensation which can increase the frequency response to MHz.

(v) Dynamic Compensation Technique

Figure 7.20 shows a scheme to use dynamic compensation network to increase the flat frequency response of the system. The whole system is

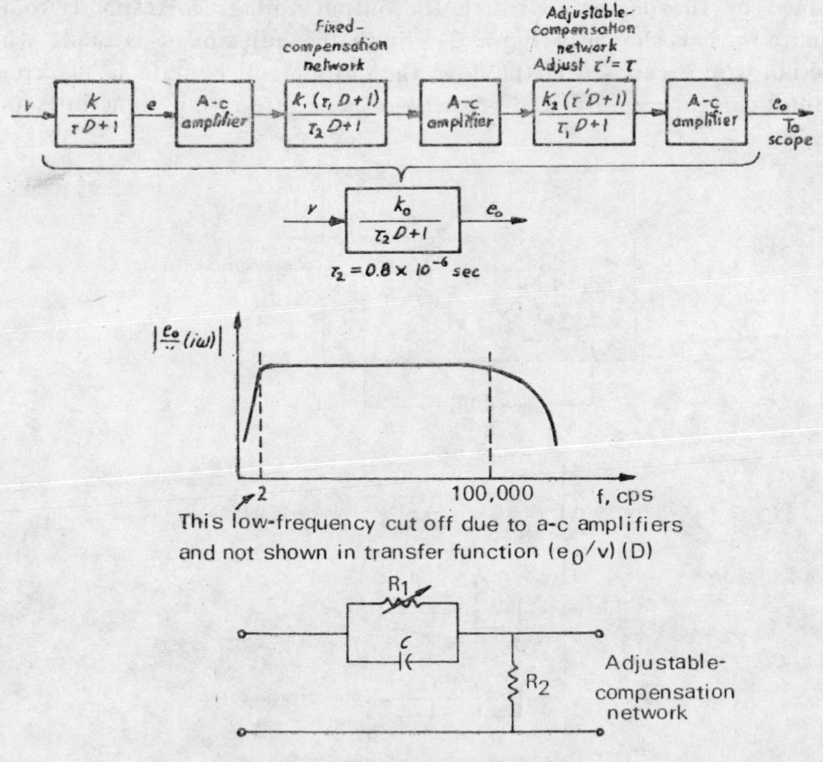

Fig. 7.20 Dynamic compensation

again equivalent to a first order system with time constant which can have a value of the order of microsecond. The frequency response of this system is shown in Fig. 7.20(b); the lower frequency cut off is mainly due to a.c. amplifier. The compensation network is a resistance shunted by a capacitor and the output is taken across the resistor. The frequency response transfer function for this network is given by

$$\frac{e_2}{e_1} = \alpha \frac{1 + i\omega\tau_c}{1 + \alpha i\omega\tau_c}$$

where $\tau_c = R_c \cdot C$

and $\alpha = R/(R + R_c)$

By the use of compensation technique the overall frequency response has been extended to a wider range, but the main difficulty is that the correct compensation depends on the value of τ itself, whose value is not known and varies with flow condition.

This difficulty can be overcome by superposing a square wave current through the hot wire while it is exposed to the flow to be studied. The output voltage response to this current signal has exactly the same time constant as the response to the flow velocity signal. Thus, if the compensation can be adjusted to be correct for the current signal, it will also be correct for the velocity signal. The correctness of the adjustment may be judged by the degree to which the output voltage corresponds to the square wave as shown in Fig. 7.21. Since the adjustment is made while the hot wire is exposed to the flow, the output will contain a superposition of current response and velocity response, resulting sometimes in a

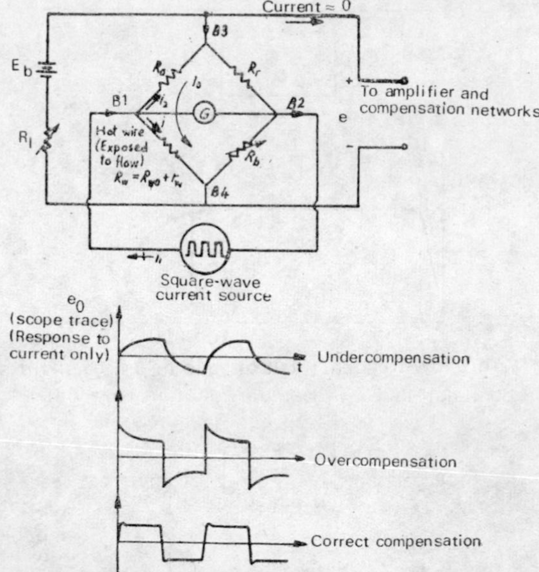

Fig. 7.21 Square wave current compensation scheme

confusing picture. Usually the compensation adjustment can be made satisfactorily.

(vi) LASER ANEMOMETER

Laser anemometer enables the measurement of the instantaneous velocity of a gas or liquid flowing in a glass walled channel. The unique features of this instrument are:

(i) Non-contact measurement,
(ii) Excellent spatial resolution,
(iii) Very fast response to fluctuating velocities,
(iv) No transfer function involvement; the output voltage is linearly proportional to the velocity, and
(v) Measurement possibilities in both gas and liquid flows.

These features establish the superiority of laser anemometer over hot wire anemometer. It is, however, restricted in usage to flows with suspensions which scatter light. The particle density of suspensions should not be less than 10^{10} particles/mm^3. The suspension size in gases ranges from 1 to 5 μm and in water from 2 to 10 μm.

The instrument is used in two modes:

(i) Reference beam mode; and
(ii) Interference fringe modes.

The advantages and disadvantages of these modes are discussed in detail later.

Consider a light beam of wavelength λ in vacuum crossing the flow stream at an angle α with the direction of flow as shown in Fig. 7.22. The frequency of light scattered from the moving suspensions is Doppler shifted with reference to the frequency of the incident light. It can be shown that if the scattered light is picked up at an angle θ with the direction of the incident light, the Doppler frequency shift is given by

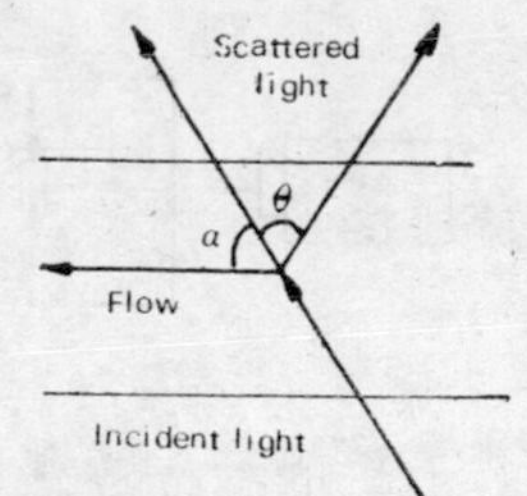

Fig. 7.22 Scattering geometry

$$\Delta f = \frac{V}{\lambda}[(\cos\theta - 1)\cos\alpha - \sin\theta\sin\alpha]$$

where v is the velocity of the suspension. In particular if

$$\alpha = \pi/2$$

$$\Delta f = \frac{v}{\lambda}\sin\theta$$

or

$$v = \Delta f \cdot \frac{1}{\sin\theta}$$

The wavelength is referred in the medium and hence

$$v = \Delta f \cdot \frac{\lambda}{n \sin \theta}$$

where n is the refractive index of the medium and λ_0 is the wavelength in vacuum.

If the incident and scattered beams are equally inclined to a normal to the flow direction, $\alpha = \pi/2 - \theta/2$ and hence

$$v = \Delta f \cdot \frac{\lambda}{2 \sin \theta/2}$$

These are the basic formulae of laser anemometry. In the reference beam method, laser light is split into two beams, which are directed at the point of measurement in the field at an angle θ. The scattered light in the direction θ is picked up and photomixed with the reference beam in the same direction in the photo multiplier tube, which yields a signal at the difference frequency, which is the Doppler shift frequency. The schematic is shown in Fig. 7.23(a). This mode of operation can be used with advantage when the suspension density is fairly high.

Further, the alignment tolerance required for the heterodyne process is not critical, thus this mode of operation is very easy to use. It is, however, to be noted that the quality of Doppler signal depends on the intensity ratio of reference and scattered beams, requiring a careful control of intensity of reference beam through the use of filters.

In the interference-fringe or differential mode the experimental arrangement is similar to that of the reference mode, as shown in Fig. 7.23(b).

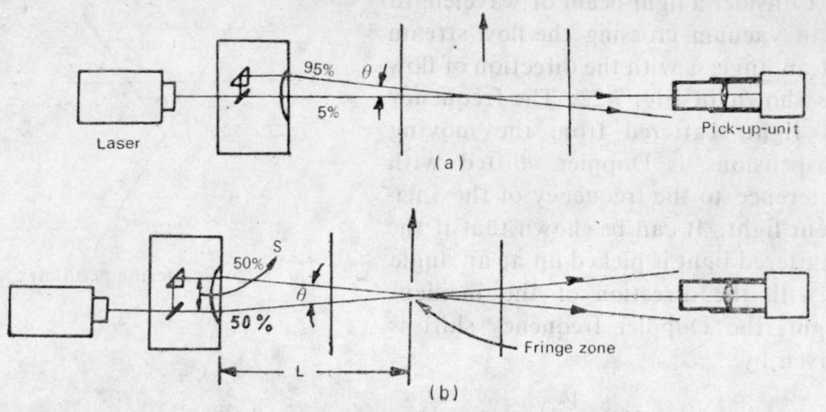

Fig. 7.23 (a) Reference beam mode
(b) Interference fringe mode

The working principle of this mode is, however, different from that of the reference mode. In the region of intersection of the two beams, Young's interference fringes are formed with spacing given by

$$\overline{x} = \frac{\lambda F}{s}$$

where s is the separation of the beam and F is the focal length of the lens which directs the beam to that point of measurement. When the suspended particles in the flow traverse the fringe pattern, they scatter light whenever they are at the bright fringe. This scattered light alone is picked up by the lens and focussed at the photo detector. Due to the random passage of innumerable particles, the output signal consists of a random amplitude modulated signal accompanied by a low frequency signal due to the traverse of the fringe pattern by the particle. From the frequency of this signal which is equivalent in magnitude to the Doppler shift frequency, the velocity of flow is obtained using the earlier expression. However, in practice a ratio of $F/s \geqslant 20$ is preferred. Because of the large volume of fringe system and larger solid angle over which scattered light is collected, this mode is applicable in medium with low suspension density. The spatial resolution of the system is poorer compared to that of reference mode.

In both these cases, using back scattering, flow measurement can be performed from the one side alone, thus requiring only one glass window in the channel. When flow velocity components are to be measured in two dimensions, the scattered light is picked up in two perpendicular planes by two independent photo detectors.

A commercial unit with flow speed range from 0.3 cm/s to 300 m/s and a good frequency response up to 150 KHz is available. Accuracy of this anemometer is 1% of full scale reading.

7.8 Measurement of mass flow rate

Some applications require the knowledge of mass flow rate rather than volume flow rate, for example, range capability of an aircraft or a rocket is determined by mass flow rate; process industries in particular chemical industries often require the knowledge of mass flow.

There are two basic approaches for metering mass. The first involves the measurement of volume flow rate with a simultaneous measurement of density. Pressure differential meters give signals proportional to ρv_{av}^2 to which ρ is multiplied and square root extracted to give mass by the processor, while the velocity flow meters give only v_{av} to which the value of density is multiplied. In the second approach some methods are used which are inherently sensitive to mass flow alone, for example; heat transfer, angular momentum transfer, coriolis acceleration, gyroscopic action, shock waves, pulsating flow, rotating pitots and vibrating pitots. The measurement of mass flow requires complex instrumentation. Several designs of mass flow meters are based on the angular momentum transfer. In one such instrument the fluid is directed axially through an impeller rotating at constant speed. The torque required to drive the impeller is a function of the mass of fluid passing through and is given by

$$T = r^2 \omega \dot{m}$$

where T is the torque, r radius at the outlet, ω the angular velocity and

$\dot{m}$ is the mass flow rate through the impeller. The torque is, therefore, directly related to the mass flow rate. It should be noted that torque will not be zero when $\dot{m} = 0$ because of the frictional effects. Furthermore, viscosity variations would also cause the torque to vary.

A variation of this approach is to drive the impeller at a constant torque with some sort of slip clutch. The impeller speed is then a measure of mass flow rate, i.e.

$$\omega = \frac{T}{r^2 \dot{m}}$$

The relationship between ω and $\dot{m}$ is inverse but the measurement of ω is far more easier compared to that of T.

7.9 Vortex shedding applied to flow metering

Vortex shedding is a phenomenon that can occur when a fluid flows past a bluff or non-streamlined body. Generally the flow does not follow the shape of the obstruction on the downstream side but separates from its surface, causing eddies to form. They grow in size until they are too large to remain attached, then break away, being shed downstream at a frequency determined by flow rate.

The direct relationship between shedding frequency and flow velocity makes the phenomenon of vortex shedding pertinent to flow metering, and a simple count of the shed vortices, called Karman vortices, is all that is required to establish total flow. The flow meter is capable of metering liquids and gases in a temperature range from $-200°C$ to $300°C$. The linearity of the instrument over a flow range of 100 : 1 is 5%.

7.10 Flow visualisation methods

The hot wire and laser anemometers are utilised for point by point mapping of the flow field; the former even disturbs the flow when it is inserted there for measurement. The flow visualisation techniques are whole field methods and provide very useful information about the flow conditions. In some cases quantitative information can also be extracted. They are divided in the following groups:

(i) Shadowgraphy,
(ii) Schlieren technique,
(iii) Interferometric methods.

These methods work due to the fact that pressure field or flow field accordingly changes the refractive index of the medium and these refractive index inhomogeneities either refract light waves or introduce a path difference when light waves propagate through it. We shall see how the path of light rays is changed as it traverses the spatially inhomogeneous continuous medium.

Consider a flow of gas as shown in Fig. 7.24. The flow is in the z

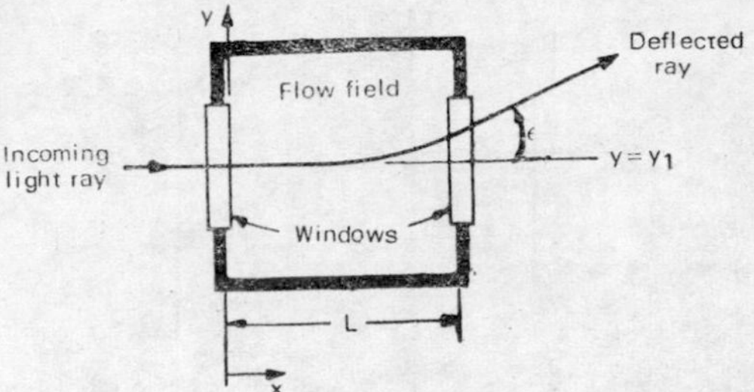

Fig. 7.24 Basic optical effects used for flow visualisation

direction, i.e., perpendicular to the figure. An incoming ray is deflected through an angle ϵ as a result of refractive index gradients in the flow. It may be shown that the deflection angle for small refractive index gradients is given by

$$\epsilon = \frac{L}{n_1}\left(\frac{dn}{dy}\right)\bigg|_{y=y_1}$$

where L is the width of the flow. The angular deflection of a light ray is directly proportional to the refractive index gradient in the flow. This is the basic optical effect on which shadowgraphy and Schlieren photography work. It may be noted that the deflection of the light ray is a measure of the average density gradient integrated over the x-coordinate. Therefore, the flow visualisation methods usually indicate refractive index variations in two dimensions only and will average the variation in the third dimension.

It is of interest to note that the refractive index for gases is related to the density through the relation

$$n = 1 + \beta\frac{\rho}{\rho_s}$$

where β is a constant (= 0.000292 for air) and ρ_s is the reference density usually taken at standard conditions. Substituting one gets

$$\epsilon = \frac{L\beta}{n_1\rho_s}\left(\frac{dp}{dy}\right)\bigg|_{y=y_1}$$

Therefore the deflection is proportional to the density gradients and hence pressure gradients can be qualitatively seen in both the techniques.

(i) SHADOWGRAPHY

Consider a passage of a collimated beam of light through a flow field. In the absence of flow, intensity distribution at the observation plane perpendicular to the direction of beam will be uniform. However, when the flow is 'on', various rays will suffer different deflections resulting in the non-uniform intensity distribution at the observation screen as shown in Fig. 7.25. The intensity at any point on the screen will depend on the relative deflection of the light rays, i.e., on

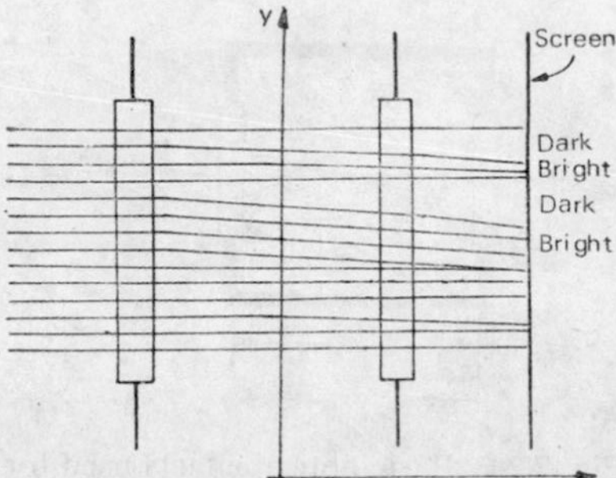

Fig. 7.25 Shadowgraph flow visualisation device

$$\frac{d\epsilon}{dy} = \frac{d^2 n}{dy^2}$$

or

$$\frac{d\epsilon}{dy} = \frac{d^2 \rho_y}{dy^2}$$

Therefore, the intensity at the screen depends on the second derivative of refractive index or density. It is almost fruitless to evaluate the refractive index or density distribution in the field. However, this is a powerful technique to show these variations and study them qualitatively. It has been utilised for the study of turbulent flow; formation and location of shock wave, etc.

(ii) SCHLIEREN TECHNIQUE

Consider a schematic shown in Fig. 7.26. A beam of light from a point

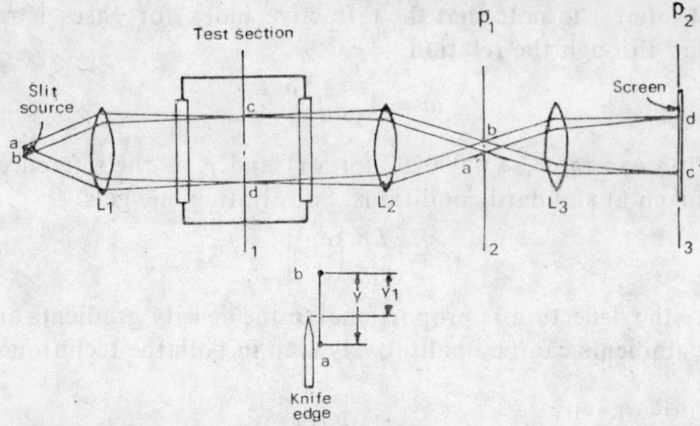

Fig. 7.26 Schematic of Schlieren method

source collimated by lens L_1 passes through the test section. This beam is focussed by a lens L_2 at a plane p_1 where a knife edge is inserted. Another lens L_3 images the test section at plane p_2. In the absence of flow, the beam passes collimated and focussed to a point at plane p_1. The influence of inserting a knife edge is to block the field uniformly resulting in a uniform movement of the cut off boundary at the plane p_2. In extreme cases, the plane p_2 is either bright or dark. However, when the flow is switched on, the rays suffer deflections and are no longer focussed to a point but scatter on a larger patch and in a non-uniform fashion. When the knife edge is inserted in the plane p_1, the rays from various regions depending on deflection are obstructed resulting in a non-uniform intensity distribution. This effect is known as *Schlieren effect*. In practice, instead of a point source a slit source is used and the knife edge is moved parallel to the width of the slit. Let the image of the slit at plane p_1 be y, of which y_1 is not intercepted by the knife edge. The intensity distribution at the plane p_2 is therefore proportional to y_1.

The angular displacement of any ray, due to the presence of flow, at plane p_1 is ϵ. Therefore, the linear displacement here will be

$$\Delta y = f \cdot \epsilon$$

where f is the focal length of the lens L_2. The intensity variations over the uniform intensity due to y_1 are due to the displacement Δy. Therefore, the contrast in the plane p_2 is

$$c = \frac{\Delta I}{I} = \frac{\Delta y}{y_1} = \frac{f_2 \epsilon}{y_1} = \frac{f_2}{y_1} \cdot \frac{L}{n_1} \left(\frac{\partial n}{\partial y} \right)$$

For gas, it can be written as

$$c = \frac{f_2}{y_1} \frac{L}{n_1} \cdot \frac{\beta}{\rho_s} \cdot \left(\frac{\partial \rho}{\partial y} \right)$$

The contrast on the screen is therefore directly proportional to the refractive index or density gradients. The contrast can be increased by decreasing y_1, i.e., intercepting more light resulting in the decrease in intensity. The contrast, therefore, cannot be reduced indefinitely and hence a compromise between contrast and illumination should be struck.

The important feature between shadowgraphy and Schlieren method is that the former senses the second derivative, while the latter the first derivative of refractive index or density. It is relatively easier to obtain quantitative information from Schlieren photographs. It will be shown that interferometry provides direct mapping of refractive index and hence best suited for quantitative analysis. However, Schlieren technique has been used for the study of turbulent flows, boundary-layer phenomena, etc.

(iii) Interferometric Techniques

In spite of the fact that there are a variety of interferometers with certain advantages, Mach-Zehnder a very simple scheme is still an unsurpassed choice of interferometer. However, it is an exceedingly difficult device to align especially for a relatively longer period. It also suffers from the disadvantage of a direct and serious influence of external influences.

The beam from a source is collimated and split into two with the help of a beam splitter B_1 (Fig. 7.27). The choice of the initial collimating

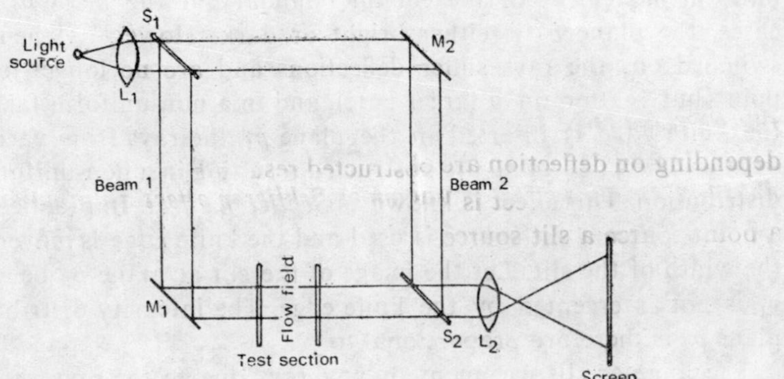

Fig. 7.27 Schematic of Mach-Zehnder interferometer

system is governed by the size of the field it is designed to cover. The two beams traverse at right angles to each other, folded by mirrors M_1 and M_2 and re-united at the beam splitter B_2, where the interference effects are observed. The flow field is in one arm of the interferometer, while the other arm has only the compensating plates. The interferometer is normally compensated for path. At least five independent motions are required for alignment and usually mirrors M_1 and M_2 are equipped with these.

The interferometer may be adjusted either to have uniform field or straight line fringes—the later is often preferred as the initial state of the system. These fringes are due to the interference between two plane wave fronts which enclose a small angle between them. The orientation of fringes can be changed by adjusting the mirror alignment. The fringes are the loci of constant optical path. Therefore, if any variation in the test field occurs, it consequently changes the optical path and thus the shape and orientation of the fringes. A quick look at the interferogram provides information regarding the regions of constant optical path changes.

The path change introduced in the beam as it passes through the test section of length L is

$$\Delta = L\beta(n - n_0)$$

where n_0 is the refractive index in the reference path. Substituting for density,

$$\Delta = L\beta \frac{\rho - \rho_s}{\rho_s}$$

The shift in the fringes will be

$$m = \frac{\Delta}{\lambda} = \frac{(n - n_0)L}{\lambda} = L\beta . \frac{\rho - \rho_s}{\rho_s \lambda}$$

where λ is the wavelength of light. When $n = n_0$, one obtains zero order fringe and therefore, the order of fringe is measured from zero order.

The zero order fringe can be easily located using white light fringes when the interferometer is compensated.

Another interesting interferometer which should find applications for flow visualisation is a triangular interferometer. This is insensitive to vibrations and environmental conditions like pressure and temperature changes because both beams follow the same path and are influenced equally. The schematic of the interferometer is shown in Fig. 7.28. The only disadvantage of the interferometer is that the reference beam also passes through the test section. The interferometer uses interference between two spherical wave fronts rather than collimated wave.

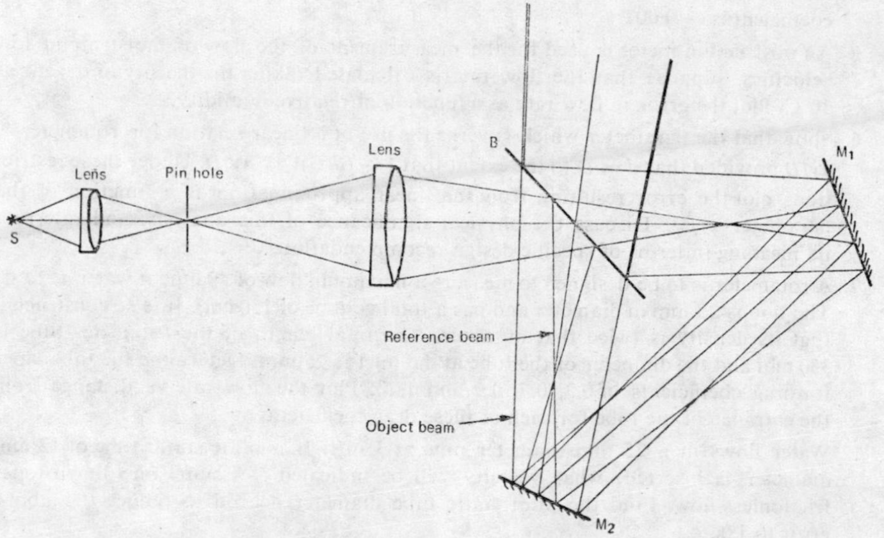

Fig. 7.28 The cyclic interferometer

The advantage of the interferometric method is that it directly gives the refractive index or density changes; these changes, however, have been integrated over the length of the test section. It finds applications to a wide range of flow conditions from the low velocity ($\simeq 0.3$ m/s) flow in the free-convection boundary layers to the shock wave phenomena in supersonic flow.

Exercises

1. A lobbed-flow meter is used for the measurement of flow of nitrogen at 1.5 N/m² and 40°C. The meter has been calibrated so that it indicates volumetric flow with an accuracy of $\pm 0.5\%$ from 30 to 100 m³. The uncertainties in the gas pressure and temperature measurements are ± 0.00015 N/m² and ± 0.5°C, respectively. Calculate the uncertainty in the mass flow measurement at the given pressure and temperature conditions.
2. Analyse the error in the flow rate measurement caused by the thermal expansion of an orifice plate.

3. A venturi tube is to be used to measure a minimum flow rate of 200 lpm at 20°C. The throat Reynolds number is to be at least 10^5 at these flow conditions. A differential gauge is selected which has an accuracy of 0.25% of full scale, and the upper limit is to be selected to correspond to the maximum flow rate. Determine the size of venturi and the maximum range of the differential pressure gauge and estimate the uncertainty in mass flow measurement at nominal flow rates of 100 and 200 lpm. The discharge coefficient is to be determined from the table.
4. An orifice with pressure taps one diameter upstream and one-half diameter downstream is installed to measure the same flow of water as in Exer. 3. For this orifice, $\beta = 0.50$. The differential pressure gauge has an accuracy of 0.25% of full scale, and the upper scale limit is selected to correspond to the maximum flow rate. Determine the range of the pressure gauge and the uncertainty in the flow rate measurement at nominal flow rates of 100 and 200 lpm. Assume that the uncertainty in flow coefficient is ± 0.002.
5. An obstruction meter is used for the measurement of the flow of moist air at low velocities. Suppose that the flow rate is calculated taking the density of dry air at 30°C. Plot the error in flow rate as a function of relative humidity.
6. Show that the parameter which governs the use of a linear relation for rotameter is $\alpha y/D$ provided that $d \simeq D$ to the extent that $1 - (d/D)^2 \ll \alpha y/D$. Under these restrictions plot the error resulting from the linear approximation as a function of the parameter $\alpha y/D$. Discuss the physical significance of this analysis, and interpret its meaning in terms of specific design recommendations.
7. A rotameter is to be designed to measure a maximum flow of 40 lpm of water at 20°C. The bob is 25 mm in diameter and has a total volume of 150 cm³. It is so constructed that its density is twice that of water. The total length of the rotameter tube is 350 mm and the diameter of the tube at the inlet is 25 mm. Determine the tube taper for drag coefficients of 0.3, 0.7, 0.9 and 1.20. Plot the flow rate vs. distance from the entrance of the tube for each of these drag coefficients.
8. Water flows in a 25 mm diameter pipe at 3 m/s. If a pitot static tube of 12 mm diameter is inserted, what velocity will be indicated ? Assume one-dimensional frictionless flow. Find the pitot static tube diameter needed to reduce the above error to 1%.
9. Water at an unknown pressure is flowing in a 50 mm diameter tube. Two simple water manometers are connected as shown in Fig. 7.29. A static pressure tube, flush with the side wall, shows 250 mm of water. The impact or pitot tube at the pipe axis shows 1000 mm of water. What is the average water pressure, and what is the flow rate $\dot{Q}$?

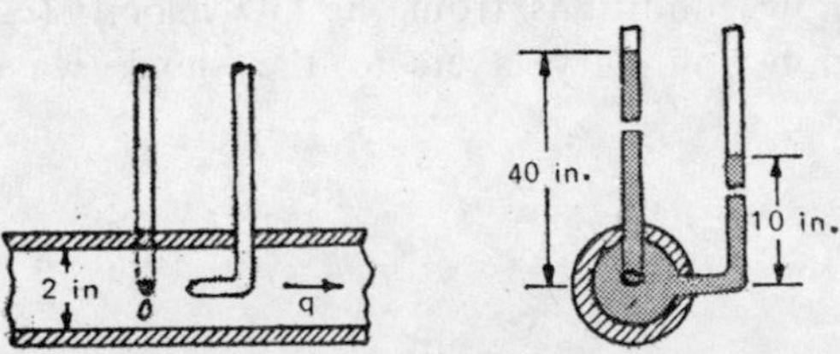

Fig. 7.29

10. Dry air at essentially atmospheric pressure and room temperature flows through a thermally insulated pipe having an area of 6 mm² as shown in Fig. 7.30. An electric heater inside the pipe dissipates 100 watts. A differential thermocouple, as shown, indicates, a temperature difference of 4°C. Estimate the average flow velocity. What are the primary sources of error in such a system?

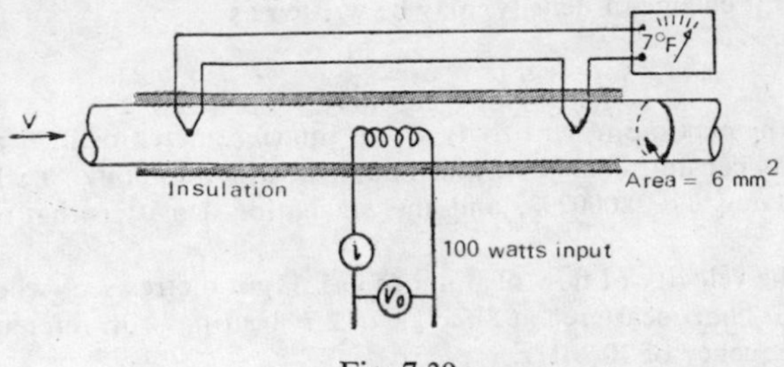

Fig. 7.30

11. A hot wire anemometer is constructed from nickel wire 25 μm in diameter and 25 mm long. It is used to measure velocity transients in air moving at an average velocity of 250 m/s, at atmospheric pressure and temperature. The wire is used as one arm of a Wheatstone bridge as shown in Fig. 7.31. The input voltage V_0 will be manually adjusted until the long time-constant meter M reads zero. The fluctuations will then be observed at the oscilloscope. Determine the expected circuit parameters (resistances and voltages) and estimate the sensitivity of the system in volts/metre/sec. (Resistivity of nickel wire 8.6×10^{-6} ohm cm. at 20°C, temp. coeff. 0.0067/°C.)

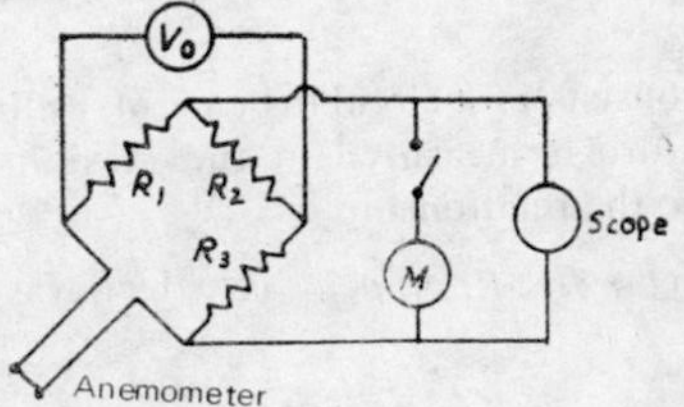

Fig. 7.31

12. Derive an expression for the product of density and velocity across a hot wire anemometer in terms of wire resistance, current through the wire, and empirical constants c_0 and c_1. Then obtain an expression for the uncertainty in this product as a function of measured quantities.
13. The sensitivity of a Schlieren system is defined as the fractional deflection obtained at the knife edge per unit angular deflection of a light ray at the test section. Show that this sensitivity may be calculated from $S = f_2/y_1$. Derive an expression for the contrast in terms of the sensitivity S, density gradient and the width of the test section.
14. An interferometer is used for the visualisation of a free convection boundary layer. For this application, the following data were collected:

Plate temperature $T_w = 50°C$
Free stream air temperature $T_\infty = 20°C$
$\beta = 0.000293$
$L = 450$ mm
$\lambda = 546$ nm

Reference density is taken at 20°C and 1.013 N/m² pressure. Calculate the number of fringes which will be viewed in the boundary layer. Calculate the temperature corresponding to the four fringes nearest to the plate surface.

15. Show that the sensitivity of an interferometer, defined as the number of fringe

shifts per unit change of density, may be written as

$$S = \frac{\beta L}{\lambda \rho_s}.$$

Show that the maximum sensitivity of an interferometer defined as the number of fringes shifts per unit change in Mach number will be about 30 when $L = 150$ mm at $\lambda = 5400$ nm, $\beta = 0.000293$, and the stagnation density is that of air at standard conditions.

16. Calculate the velocity of flow of water ($n = 1.33$) in metres/sec. when a beam from laser ($\lambda = 632$ nm) scattered at an angle of 2.7° beating with reference wave, gives a Doppler frequency of 10 MHz.
17. A spherical ball is suspended by a fine string from a fixed point as shown in Fig. 7.32. The wind moving at velocity V causes the string to be at an angle ϕ relative to vertical. Determine a relation between ϕ (degrees) and v (km/h).

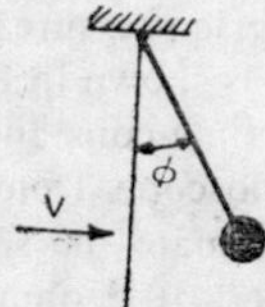

Fig. 7.32

18. A liquid flow meter consists of a circular bend of radius R in a pipe as shown in Fig. 7.33. Static pressure is measured in the straight section p_1 and in the curved section p_2. Determine the relationship

$$\dot{Q} = f(\rho, R, p_1, p_2) \quad \text{(Neglect gravity)}$$

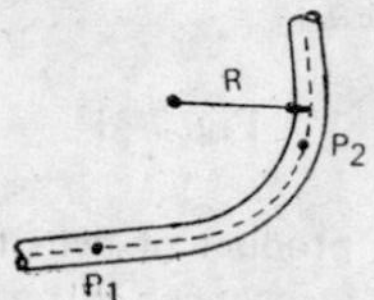

Fig. 7.33

What are the probable sources of error?

8

Measurement of Temperature

8.1 Introduction

The concept of temperature is a very difficult one. Temperature is usually defined as a measure of 'degree of heat' while heat is taken to mean the quantity of heat. It is equivalent to potential in electricity and level in hydrostatics. Higher temperature would mean that the heat would flow from it to the lower temperature irrespective of the heat content of the body at the higher temperature. The temperature can, therefore, be said to be 'heat level'.

On the basis of kinetic theory, the temperature of a system may be defined through the equation

$$\tfrac{1}{2}mv_{av}^2 = kT$$

where m is mass of the molecule, v_{av} its average velocity and k is the Boltzmann's constant. The equation holds good for systems which obey Maxwell-Boltzmann distribution, i.e., for gases. However, for liquid and solid systems, the temperature may again be defined by the intensity of molecular activity and the relation is very complicated due to the presence of strong inter-molecular forces. Therefore, temperature may be considered as a manifestation of the molecular kinetic energy of the body. All motions stop at absolute zero of temperature. There are, however, other definitions of the temperature also which have been proposed by various researchers. Two of them are:

(i) A condition of a body by virtue of which heat is transferred to or from other bodies.

(ii) A quantity whose difference is proportional to the work from a Carnot engine operating between a hot source and a cold receiver.

Since temperature is a derived quantity, its definition is very difficult. Also, no such standard as the standards of mass, length, etc., can be defined. The temperature is thus measured by the measurement of certain properties of matter which are influenced by the degree of heat. The most used are changes in:

(i) physical state,
(ii) chemical state,

(iii) dimensions,
(iv) electrical properties,
(v) radiation properties.

None of these changes gives an absolute measure of temperature but must be calibrated to some arbitrarily chosen standard.

8.2 Expression of temperature

The temperature is expressed by degree centigrade, Fahrenheit or Rankine. All these scales depend on the properties of the substance. In fact, on Centigrade scale the difference between the freezing point and boiling point of water at NTP has been divided into 100 equal parts; each is called as one degree Centigrade. The absolute temperature scale proposed by Lord Kelvin does not depend on the properties of the substance used and makes the basis of all thermodynamical quantities.

The international temperature scale of 1948 sets the basis of an experimental scale and has some well defined standard points as close as possible to the thermodynamic scale. The following basic fixed points were adopted, all corresponding to normal atmospheric pressure of 760 mm of Hg, and are used to standardise thermometers in the laboratory. They are:

(i) Oxygen point—boiling point of liquid oxygen (−182.97°C),
(ii) Ice point—melting point of pure ice (fundamental point) (0°C),
(iii) Steam point—boiling point of pure water (fundamental point) (100°C),
(iv) Sulphur point—boiling point of liquid sulphur (444.60°C),
(v) Silver point—melting point of silver (960.8°C),
(vi) Gold point—melting point of gold (1063.0°C).

The points on the scale between the fixed points are interpolated as follows:

1. From −190°C to 0°C—By the measurement of resistance of a standard platinum resistance thermometer. The resistance, R_t, at any temperature, t, within this range is given by

$$R_t = R_0[1 + At + Bt^2 + C(t - 100)t^3]$$

where the constants R_0, A, B and C are obtained by measuring the resistance at the ice point, steam point, sulphur point and oxygen point.

2. From 0°C to 630.5°C—By the measurement of resistance of a standard platinum resistance thermometer. The resistance, R_t, at any temperature, t, between this range is given by

$$R_t = R_0[1 + At + Bt^2]$$

where the constants R_0, A and B are obtained by measuring resistances at ice, steam and sulphur points.

3. From 630.5°C to 1063.0°C—By the measurement of emf, e, of a

standard platinum, platinum-rhodium thermocouple; one junction being kept at the ice point, while the other junction at the temperature 't' °C. The emf is given by

$$e = a + bt + ct^2$$

where the constants a, b and c are obtained by measuring emf's at the freezing point of antimony (630.5°C), silver point and gold point.

4. Above 1063.0°C—The temperature of a body above the gold point is measured by comparing the intensity of radiation of a particular wavelength emitted by the body, with the radiation emitted by the black body at the gold point.

For temperatures below the oxygen point, the international practical temperature scale is not defined. However, the National Bureau of Standards has standardised thermometers up to 2°K, although they are not internationally accepted. They include the accoustical interferometer, the helium-vapour thermometer, and the platinum resistance thermometer.

Apart from the primary standard points described above, there are quite a few secondary points in the international scale of 1948. Some of them are given below:

(i) Temperature at which solid carbon dioxide changes to gaseous CO_2 at normal atmospheric pressure (−78.50°C),
(ii) Freezing point of mercury (−38.87°C),
(iii) Freezing point of antimony (+630.5°C),
(iv) Freezing point of palladium (+1552°C),
(v) Melting point of tungsten (+3380°C),

The approximate range and accuracy of various temperature measuring devices has been given in Table 8.1.

TABLE 8.1

No.	Type	Range °C	Accuracy °C
1.	Glass thermometers		
	(i) mercury-filled	−39 to 400	0.3 to 1
	(ii) ($Hg + N_2$)-filled	−39 to 540	0.3 to 5.5
	(iii) alcohol-filled	−70 to 65	0.5 to 1
2.	Pressure-gauge thermometers		
	(i) vapour pressure type	11 to 200	1 to 5.5
	(ii) liquid or gas-filled	−130 to 540	1 to 5.5
3.	Bimetallic thermometers	−74 to 540	0.3 to 14
4.	Thermocouples		
	(i) base metals	−185 to 1,150	0.3 to 11
	(ii) precious metals	−185 to 1,150	0.3 to 11
5.	Resistance thermometer	−240 to 980	0.003 to 3
6.	Thermistors	−100 to 260	Depends on ageing
7.	Pyrometers		
	(i) optical	760 and above	11 for black body
	(ii) radiation	540 and above	11–16 for black body
8.	Fusion	590 to 3600	As low as 20–30 under optimum conditions.

The instruments for measuring temperature have been classified, in the first place, according to the nature of change produced in the testing body by the change of temperature. The following four broad categories have, therefore, been proposed:

(i) Expansion thermometers,
(ii) Change of state thermometers,
(iii) Electrical methods of measuring temperature,
(iv) Radiation and optical pyrometry.

8.3 Expansion thermometers

The thermometers falling under this category have been further sub-divided as follows:

1. Expansion of solids
   (a) Solid-rod thermostats
   (b) Bimetallic thermostats/thermometers
2. Expansion of liquids
   (a) Liquid-in-glass thermometer
   (b) Liquid-in-metal thermometer
3. Expansion of gases
   (a) Gas thermometers

1. Expansion of Solids

(a) Solid-rod thermostats/thermometer

A temperature controlling device may be designed using the principle that some metals expand more than others for the same range of temperature. They, thus, make use of differential expansion.

The temperature sensitive portion of the instrument consists of an invar (nickel-steel) rod encased in a brass tube. The lower end of the invar rod is hard soldered to the containing tube, when this combination or stem is heated, the brass tube ($\alpha = 34.2 \times 10^{-6}/°C$) expands more than the invar rod ($\alpha = 2.7 \times 10^{-6}/°C$) so that the position of the free end of the rod changes with respect to the end of the tube. This change in the relative position is used to operate a micro-switch. The position of the switch relative to the invar rod can be adjusted by means of a knob, so that the supply of electricity is cut off when the temperature of thermostat has reached a pre-set temperature. This kind of thermostat is used in water heaters and domestic electric ovens.

(b) Bimetallic thermostat/thermometer

When the two metal strips having different coefficients of expansion are brazed together, a change in temperature causes a free deflection of the assembly as shown in Fig. 8.1. When the temperature to be measured is higher than the brazing temperature, it bends towards the low coefficient side, while if the temperature is less than the bonding temperature, it

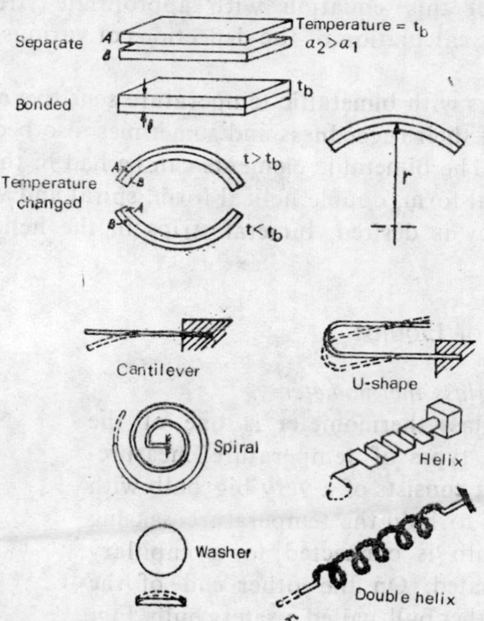

Fig. 8.1 Bimetallic sensors

bends in the other direction. Such bimetallic strips are the basis for many control devices and are also used to some extent for the temperature measurement. The following relation may be used to determine the radius of curvature of such a strip which is initially flat at the temperature t_b.

$$r = \frac{s\left[3(1+m)^2 + (1+mn)\left(m^2 + \frac{1}{mn}\right)\right]}{6(\alpha_2 - \alpha_1)(t - t_b)(1+m)^2}$$

where

s = combined thickness of the bonded strip,

m = ratio of the thickness of low expansion to that of the high expansion material,

n = ratio of modulus of electricity of low expansion to that of the high expansion material,

α_1 = linear expansion coefficient of low expansion material,

α_2 = linear expansion coefficient of high expansion material,

t = temperature,

t_b = bonding temperature.

In most practical cases $m \simeq 1$

and $n \simeq 1$

Thus the above equation reduces to

$$r \simeq \frac{2s}{3(\alpha_2 - \alpha_1)(t - t_b)}$$

Combination of this equation with appropriate strength-of-materials relations allows calculation of the deflections of various types of elements in practical use.

Thermometers with bimetallic temperature sensitive elements are often used because of their ruggedness and sometimes also because of their convenient shape. The bimetallic elements can be had in the cantilever form, U-shape, helical form, double helical form, spiral and waser form. When greater accuracy is desired, bimetal strips in the helical or spiral form are used.

2. Expansion of Liquids

(a) *Liquid-in-glass thermometer*

The liquid-in-glass thermometer is one of the most common types of temperature measurement devices. It consists of a very big bulb with very thin walls to hold the temperature sensing liquid. The bulb is connected to a capillary which is graduated. On the other end of the capillary is another bulb called a safety bulb. Fig. 8.2 shows a sketch of a thermometer. Mercury and alcohol are most commonly used liquids. The thermometer works on the principle of differential expansion of liquid. The following points may be borne in mind while designing a thermometer:

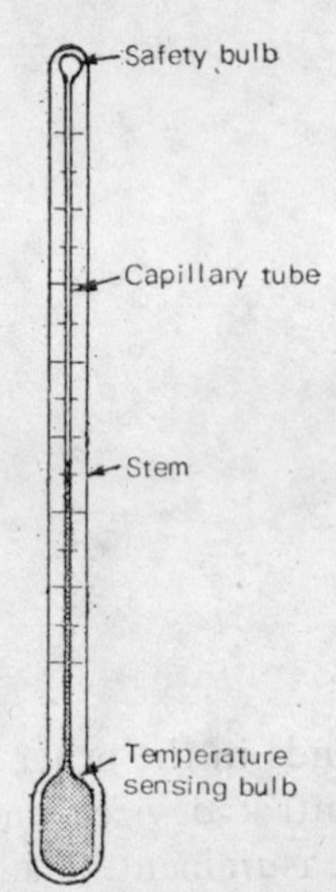

Fig. 8.2 Schematic of a mercury-in-glass thermometer

(i) Walls of liquid bulb should be thin, thereby quick transfer of heat is possible and hence the fast response provided the volume of the liquid is small. For higher sensitivity, the volume of liquid should be large, thus the response is impaired. A compromise is, therefore, to be made between sensitivity and response.

(ii) Capacity of bulb is many times the volume of capillary and hence a safety bulb at the other end is provided.

(iii) Alcohol may be preferred over mercury due to its higher coefficient of expansion.

Mercury-in-glass thermometers are generally employed up to 340°C (boiling point of mercury is 357°C) but their range may be extended to 560°C by filling the space above mercury with CO_2 or N_2 at high pressure, thereby increasing its boiling point and hence the range. The accuracy of these thermometers under optimum conditions does not exceed 0.1°C. However, when an increased accuracy is required, Beckmann thermometer can be used. It contains a big bulb attached to a very fine capillary. The range of the thermometer is limited to only 5 to 6°C with an accuracy of 0.005°C. The thermometers are used in two ways—complete immersion, and partial immersion.

The *complete immersion thermometers* are calibrated to read correctly when the liquid column is completely immersed in the temperature medium. Since this obscures reading of the scale, a little portion of the liquid column is allowed to project a little to permit observation.

The *partial immersion thermometers* are immersed to a definite liquid column and the exposed portion is to be at a temperature for which calibration is valid. However, if the temperature of the stem is different from that used for calibration, proper correction can be applied as follows:

$$\text{correction} = 0.00016(t_1 - t_2)n \quad °\text{C scale},$$
$$= 0.00009(t_1 - t_2)n \quad °\text{F scale},$$

where t_1 is the temperature of the stem, t_2 is the calibration temperature and n is the number of degrees exposed. The numerical value is the coefficient of apparant expansion of mercury in glass.

(b) *Liquid-in-metal thermometers*

MERCURY-IN-STEEL THERMOMETER. Two distinct disadvantages of liquid-in-glass thermometers are:

(i) glass is very fragile and hence immense care should be exercised in handling these thermometers, and
(ii) the position of thermometer for accurate temperature measurement is not always the best position for reading the scale of the thermometer.

Both of these disadvantages are overcome in mercury-in-steel thermometer. The principle of operation is again differential expansion of liquid which is used in a modified form as explained below:

Since metals are opaque, i.e., mercury column is not visible, the change in volume due to temperature change is thus obtained via the pressure change which is read off by the Bourdon tube (Fig. 8.3). The thermometer

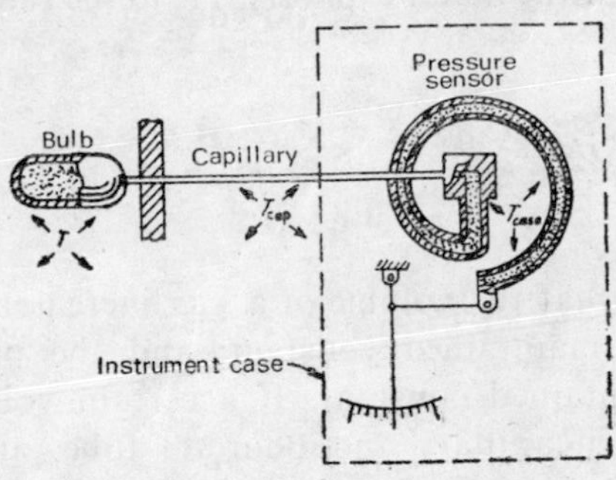

Fig. 8.3 Pressure thermometer

bulb, capillary and Bourdon tube are filled with mercury, usually at high pressure. When the thermometer is used for the measurement of temperature change, the volume of mercury increases, which exerts a force (pressure) on the Bourdon which starts uncurling. For easy and safe operation, a very long pressure transmitting system, i.e., capillary can be taken.

Various forms of Bourdon tubes are used to increase sensitivity; often helical or spiral forms of Bourdon tubes are used. Since capillary is very long, the effect of ambient temperature may be very serious and require compensation. Fig. 8.4 shows two temperature compensated arrangements: in one spiral Bourdon tube and in other C-type Bourdon tube is used.

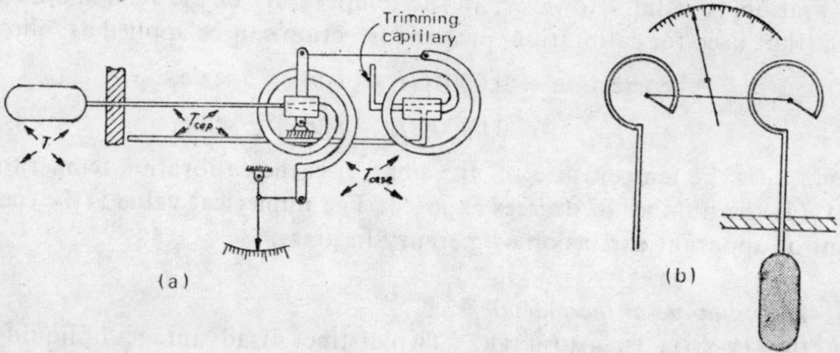

Fig. 8.4 Compensation methods

The length of the capillary in the compensating Bourdon tube is equal to that of measuring thermometer and runs parallel to it. Another method is offered by Taylor Instruments Co. for ambient temperature compensation, which is called the Accuratus mercury actuated tube system. The system contains a filler wire in the stainless steel Accuratus tubing. The filler wire extends throughout the bore of the tubing. This wire is made of invar metal (low temperature coefficient). It decreases the volume of mercury in the capillary to the point where the volumetric expansion of mercury with a given increase in the ambient temperature exactly equals the increase in volume of the metal capillary due to the same ambient temperature increase.

3. Expansion of Gases

(a) *Gas thermometer*

It is well known that the volume of a gas increases with temperature, if the pressure is maintained constant; and the pressure increases if the volume is maintained constant. If a certain volume of inert gas is enclosed in a bulb, a capillary and Bourdon tube, and most of the gas is in the bulb, then the pressure as indicated by the Bourdon tube may be calibrated in terms of the temperature of the bulb. This is the principle of gas-filled thermometers.

This small change in pressure per degree change in temperature also limits the actuating power to Bourdon gauge because of low pressure change per degree change in temperature, and is used only to move the pen rather than for temperature compensation, where compensating

Bourdon tube is to be driven. Thus these thermometers are compensated only for case. Further the bulbs of these thermometers are made large so that the volume of gas in the bulb is very large compared to that of capillary, thus reducing the effect of ambient temperature. But this reduces the response of thermometer for dynamic changes. But the response is faster compared to liquid thermometers of the similar bulb size.

8.4 Change-of-state thermometers

Let us consider a container (Fig. 8.5) having a certain quantity of liquid; the space above it is assumed to be empty. The molecules of liquid are always in the state of random motion, moving in all the directions with different velocities. If a molecule has a vertical component of kinetic energy greater than the force of attraction at the liquid surface, it escapes from the liquid. Thus there would be a stream of liquid molecules escaping. This constitutes vapour. On the other hand, the molecules which constitute vapour are also moving at random and a few are attracted by the liquid. Thus the processes of evaporation and condensation go on simultaneously. When the rates of evaporation and condensation are equal, the vapour becomes saturated.

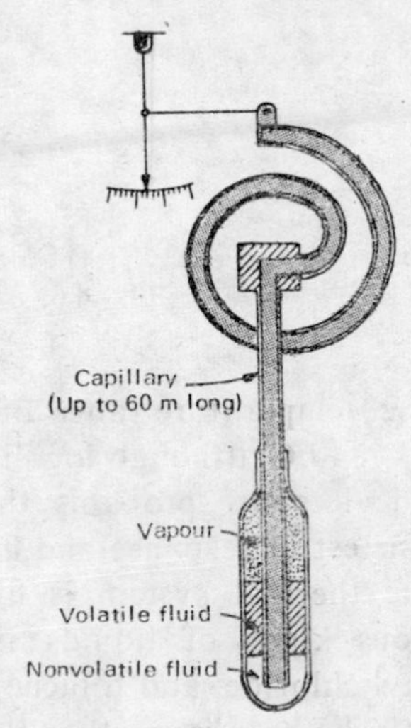

Fig. 8.5 Vapour pressure thermometer

It is to be proved now that the vapour is always saturated when in contact with the liquid. If the size of the upper part of the container is alterable, the vapour pressure will fall momentarily with an increase in volume and rate of evaporation would increase till the vapour is saturated and vice versa. Thus the saturated vapour pressure depends only on the temperature and properties of the liquid and is independent of the size of the container. A typical graph illustrating the variation of saturated vapour pressure of water with temperature is shown in Fig. 8.6. The relationship is non-linear.

Thus, provided there is always liquid and vapour present, the saturated vapour pressure of the liquid depends only on the temperature, and is independent of the size of the container. Further, if a thermometer system similar to that described for gas expansion thermometers is arranged so that the system contains both liquid and vapour and the interface between the liquid and vapour, is in the bulb (that is at the temperature whose value is to be determined), then the vapour pressure as measured by the Bourdon tube would give an indication of the temperature.

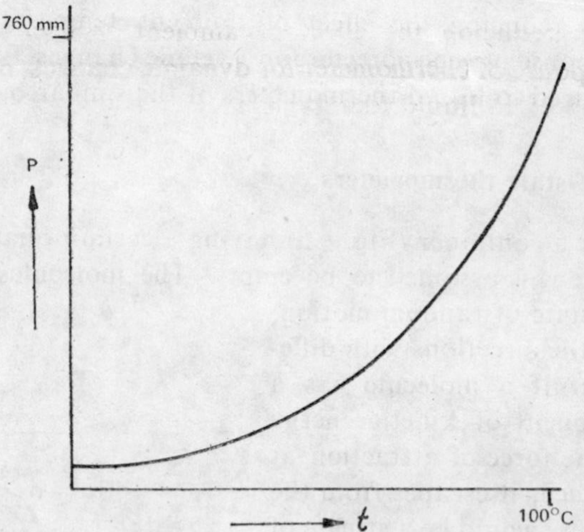

Fig. 8.6 Variation of vapour pressure of water with temperature

The temperature range of these systems is generally between −40°C and +340°C although low limits of −185°C are available. These measuring devices are probably the most widely used thermal systems, being the fastest in response and lowest in cost.

The thermal system is filled with a volatile liquid and its vapour. Various kinds of liquids like propane, sulphur-dioxide, ethyl ether, methyl chloride, and toluene are used depending on the range of instrument. The bulb is partly filled with the liquid, the rest of the thermal system being filled with the same material in its vapour form.

The accuracy of vapour pressure thermometers is not affected by changes in the ambient temperature as long as these changes do not oscillate around the process temperature, which the instrument is to measure. Suppose the process temperature is between 15.5°C and 18.3°C and the ambient temperature around the capillary and Bourdon tube fluctuates between 10°C and 25°C. When the ambient temperature cuts above and below the process temperature, a temporary unbalance will occur. If, for example, a rapid increase takes place in the ambient temperature from 10°C to 25°C, the vapour in the capillary and Bourdon tube will expand and the internal pressure will increase, causing the pen to move upscale, although the process temperature has not changed. However, the condensation of vapour will take place till the equilibrium state is achieved and the pen will return to the correct reading as determined by the temperature of the bulb.

The relationship between pressure and temperature in change-of-state thermometers is not linear; the result is an instrument scale with spacing that gradually increases with temperature. However, by using a linkage between measuring Bourdon tube and the pen, a non-linear response of the instrument is changed into a linear pen movement. In other cases, the range is limited and only $\frac{1}{3}$ of the scale can be effectively used. However, by suitable choice of filling liquids, wide variety of ranges may be available. The choice of material for bulb construction is also very wide. Metals and alloys such as copper, steel, monel, tantalum, etc., may be used. These may be plated with nickel, chromium or silver, if necessary. The advantages and limitations of these thermometers are given below:

(i) They are cheaper than mercury-in-steel thermometers,
(ii) They respond faster,
(iii) Do not suffer from ambient temperature effects,
(iv) They can be constructed with smaller bulb than the other types,
(v) Have limited range due to non-linear scale,
(vi) The range of the instrument is limited by the fact that the minimum temperature for which it can be used must be well below the critical temperature.

8.5 Pyrometric cones

At certain definite conditions of purity and pressure, substances change their state at fixed temperatures. This fact forms a useful basis for fixing the temperatures.

The melting points of certain minerals and their mixtures are used to find out the temperature of a kiln in the ceramic industry. The mixtures which consist of silicate minerals such as Kaolin (aluminium silicate), Talc (magnesium silicate), Felspar (sodium aluminium silicate), quartz (silica), etc., along with other minerals such as calcium carbonate are made up in the form of cones known as Seger cones. By varying the composition of cones a range of temperatures from 600 to 2000°C may be covered in steps of 25 or 45°C. A series of cones is placed in the kiln. Those of lesser melting point will melt, but eventually a cone will be found which will just bend over. This cone indicates the temperature of the kiln. The maximum sensitivity is ± 10°C.

Other kinds of items serve to indicate by change of colour or appearance when a certain temperature has been exceeded. They are available in 7 to 25°C steps from 45 to 1370°C. They are valuable for indicating excessive temperature as encountered in electrical equipment, retorts, and piping.

8.6 Electrical methods

In electrical methods of measuring temperature, the temperature signal is converted into electrical signal either through a change in resistance or voltage, leading to a change in current development of emf. The following elements are used to convert temperature into electrical variables:

1. Electrical resistance bulb,

2. Thermistors, and
3. Thermocouples and thermopiles.

1. Electrical Resistance Bulb

Most of the work in connection with electrical resistance thermometry involves the knowledge of Ohm's law. It is found that the resistance of pure metallic conductors increases with temperature. In practice, platinum, copper and nickel are used as resistance elements because they can be obtained in a high degree of purity and possess a high degree of reproducibilities of resistance characteristics.

These elements can be used in the following ranges:

Platinum	$-190°C$ to $+630.5°C$,
Copper	$-50°C$ to $+250°C$,
Nickel	$-200°C$ to $+350°C$.

At temperatures approaching absolute zero, leaded phosphor bronze resistance thermometers are found to be more suitable. Thermometers using platinum as the resistance element are discussed here.

Platinum resistance thermometer

Platinum is the standard used in the resistance thermometer that defines the International Temperature Scale of 1948, not because it has particularly high temperature coefficient but because of its stability in use. The presence of impurities in platinum is undesirable. The temperature coefficient of resistance is also sensitive to internal strains. Hence it is essential that the platinum should be annealed at a temperature higher than the maximum temperature of service. The combination of purity and adequate annealing is shown by the ratio of resistances at steam and ice points. The accepted International Temperature Scale of 1948 value is 1.3910 or higher.

Platinum scale

If the temperature can be regarded as bearing a constant relationship to resistance, the temperature may be expressed in terms of platinum scale of temperature. This scale approximates the thermodynamic scale. If the rate of resistance change with temperature is constant between 0 and 100°C, the resistance R_t at temperature t within this range is given in terms of resistance R_0, at 0°C, by the equation

$$R_t = R_0(1 + \alpha t)$$

where the temperature coefficient of resistance α, in terms of resistance at 0°C and 100°C, is given by

$$\alpha = \frac{R_{100} - R_0}{100 R_0}$$

The value of R_0 and α may be found by measuring the resistance values of the wire at ice point 0°C and steam point 100°C. The difference between the resistance values at 0°C and 100°C is called the 'fundamental interval' for the thermometer.

The temperature t_p as measured on the platinum resistance scale may be determined by measuring the resistance R_t at the required temperature when:

$$t_p = \frac{R_t - R_0}{R_{100} - R_0} \times 100$$

The temperatures above 100°C as measured on the platinum scale will be less than those given by the gas scale or international scale. In the range of 0°–630.5°C, the resistance is best represented by

$$R_t = R_0(1 + At + Bt^2)$$

The value of third constant B is determined by measuring the temperature at the sulphur point 444.6°C. If the simple equation, $R_t = R_0(1 + \alpha t_p)$, is used for the measurement of temperature above 100°C, the value of temperature on the platinum scale will be less than that given on the international scale. The correction for this is obtained as follows:

The equation which relates the temperature at the international scale with the resistance is

$$R_t = R_0[1 + At_I + Bt_I^2]$$

and the temperature at the platinum scale with the resistance is

$$R_t = R_0[1 + \alpha t_p], \quad \text{with } \alpha = \frac{R_{100} - R_0}{100R_0}$$

Up to 100°C, both these equations give identical results, i.e.,

$$R_0[1 + 100A + 100^2B] = R_0[1 + 100\alpha]$$

At any other temperature above 100°C,

$$R_t - R_0 = R_0[At_I + Bt_I^2]$$

Substituting for $R_t - R_0$ from equation $R_t = R_0(1 + \alpha t_p)$,

$$\alpha \cdot t_p = At_I + Bt_I^2$$

or

$$t_p = \frac{At_I + Bt_I^2}{A + 100B}$$

This equation can be rewritten as

$$t_I - t_p = \delta[(t_I/100)^2 - (t_I/100)]$$

where $\delta = -\dfrac{100^2B}{A + 100B}$.

This is known as Callender formula and holds good over a large temperature range above 100°C. The temperature in international scale from platinum scale is obtained by the method of successive approximation. The value of constant δ depends on the purity of resistance element, and for pure platinum its value is 1.494. A similar equation which holds good between 0°C and −200°C can also be derived. It is known that the resistance, in this range, is given by

$$R_t = R_0[1 + At_I + Bt_I^2 + C(t_I - 100)^3]$$

The difference between the temperatures in the international and platinum scales can be shown to be

$$t_I - t_p = \delta\left[\frac{t_I}{100} - 1\right]\frac{t_I}{100} + \beta\left[\frac{t_I}{100} - 1\right]\left[\frac{t_I}{100}\right]$$

where β is another constant whose value depends on A and B. This equation is known as Callender-van Dusen equation.

Resistance thermometer bulb

Resistance thermometer bulbs are either tip-sensitive or stem-sensitive. The tip-sensitive thermometers are used for the measurement of surface temperatures while stem-sensitive are used for other applications.

In modern stem-sensitive thermometers, the coil consists of pure platinum wire wound on a former of mica, stealite, or porcelain so that it is completely free from strain. Fig. 8.7 is a cut-section of the stem of a thermometer. A spring provides a resilient cushion against the vibration as well as close thermal contact with the shell on one side and with the sensitive winding through the mica insulation on the other. After winding the resistance element is annealed at a temperature above the maximum temperature of service. The annealing relieves stresses and ensures the reproducibility of resistance readings; but care must be taken to prevent the wire becoming contaminated during the annealing process. In laboratory, it is often wound on mica frame and has a fundamental interval of 1 Ω or 10 Ω and can be used in the range of −200 to +600°C. The resistance of bulb ranges from 10 ohms to as high as 25,000 ohms. Higher resistance elements are less affected by lead wire and contact resistance variations, and since they generally produce large output voltage signals, spurious thermoelectric emf's due to joining of two dissimilar metals are also usually negligible. In other cases, the ends of platinum coil are usually connected to heavy copper or constantan leads by means of gold wires. Gold is used because it has a low resistance, a low thermoelectric emf against copper, and is easily workable. The leads are insulated from each other and from the sheath by means of beads or discs of porcelain. All possible precautions are taken to avoid contact resistance, and thermo-electric emf by arranging the junctions at the same temperature. In thermometers for measuring temperatures above 500°C, the platinum coil is connected to platinum leads.

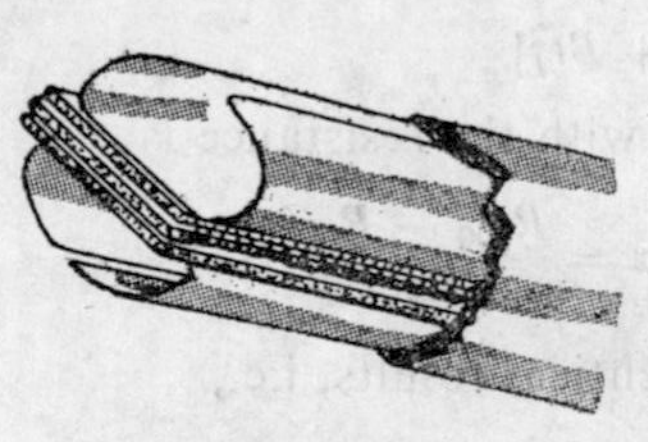

Fig. 8.7

Measurement of resistance

To measure the temperature, resistance of the coil is measured by some form of the bridge. The bridge can be used either in the null position or deflection position. Null method, however, gives better results. Since sensing element is some distance away, copper or constantan cables are used to make connections. The cross section of these cables should be so chosen as to have the resistance of the leads very small compared to that of the resistance element. Further, any variation of resistance due to ambient temperature should be compensated for. The following methods are often used:

(i) THREE-LEAD ARRANGEMENT This was introduced by Siemens in 1871 and is shown in Fig. 8.8 (a). Three leads l_1, l_2, l_3 are connected to the resistance element and the bridge circuitry. The lead l_1 is in the same arm of the bridge as the resistance R while l_3 is in the thermometer arm.

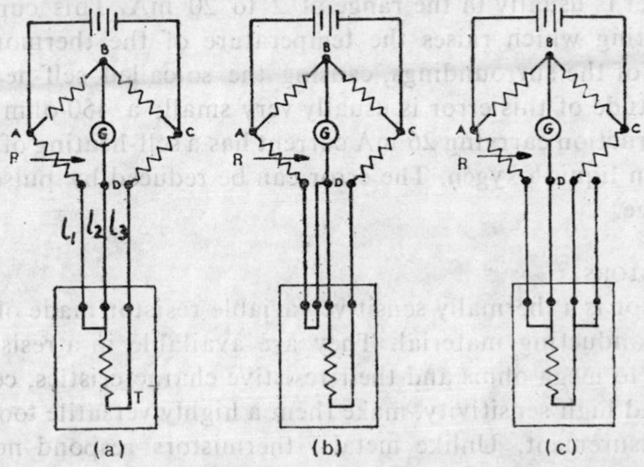

Fig. 8.8 Three methods for compensating lead resistance

The lead l_2 connects the bridge point D to one terminal of the element. It will be seen that if l_1 and l_3 are identical wires and have the same length and cross section, they will have equal resistances, and any change of ambient temperature will affect them equally, thus the bridge always remains balanced. Further, when the bridge is balanced, no current flows through the lead l_2. This may be considered an advantage. The effect of the variable and unknown contact resistance in the adjustable resistor has no influence on the resistance of the bridge arm at null balance.

(ii) CALLENDER'S METHOD The arrangement of connections is shown in Fig. 8.8 (b). Two identical pairs of leads are connected in arms AD and CD of the bridge. The pair in the arm AD are connected together at a point near the thermometer bulb, while the second pair which are in the arm CD, are connected to the thermometer bulb. Both pairs of leads are enclosed in the same outer cover so that the leads in AD compensate for any changes in the resistance of the leads in CD due to ambient temperature variations. This method is quite useful when thermometers are used in both arms to measure differential temperature.

(iii) FOUR-LEAD ARRANGEMENT This arrangement is shown in Fig. 8.8 (c) and is used in the same way as the one with three leads. Provision is made, however, for using any combination of three, thereby checking for unequal lead resistance. By averaging the readings more accurate results are possible. Some form of this arrangement is used where highest accuracy is required.

The principle of the Wheatstone bridge allows the replacement of resistances by inductances and capacitances. In one variant of resistance thermometer, made by Foxboro Co., two arms of the bridge are formed by two capacitors, one of them being the balancing capacitor. This reduces the influence of contact resistance.

The thermometers are excited by a d.c. or a.c. voltage; the current in the thermometer is usually in the range of 2 to 20 mA. This current causes an I^2R heating which raises the temperature of the thermometer bulb, above that of the surroundings, causing the so-called self-heating error. The magnitude of this error is usually very small; a 450-ohm element in open construction carrying 26 mA current has a self-heating of 0.1°C when immersed in liquid oxygen. The error can be reduced by pulse excitation of the bridge.

2. THERMISTORS

Thermistor is a thermally sensitive variable resistor made of a ceramic-like semi-conducting material. They are available in a resistance range from ohms to mega ohms and their resistive characteristics, coupled with stability and high sensitivity, make them a highly versatile tool for temperature measurement. Unlike metals, thermistors respond negatively to temperature, and their temperature coefficient of resistance is about 10 times higher than that of platinum or copper. Fig. 8.9 shows resistance temperature characteristics of thermistors.

Thermistors are made of metal oxides and their mixtures, namely, oxides of cobalt, copper, nickel, manganese, iron, tin, magnesium, titanium, uranium and zinc. The oxides, usually compressed into desired shape from powders, are heat treated to recrystallize them, resulting in a dense ceramic body. Electric contact is made by various means: wires imbedded before firing the material, plating, or metal ceramic coatings backed on. They are available in various forms like beads as small as 0.4 mm in diameter, discs ranging from 5 to 25 mm in diameter, and rods from 0.8 to 6.00 mm diameter and up to 50 mm in length (Fig. 8.10). Flakes (only a few micrometers thick) are employed as infrared radiation detectors or bolometers. Size and shape of a thermistor depend on several factors, such as space available, the required speed of response to temperature changes and the amount of power dissipation. Various methods of mountings are used: beads are suspended from wire leads or imbedded in probes; discs are mounted on spring-loaded stacks with or without heat dissipating fins;

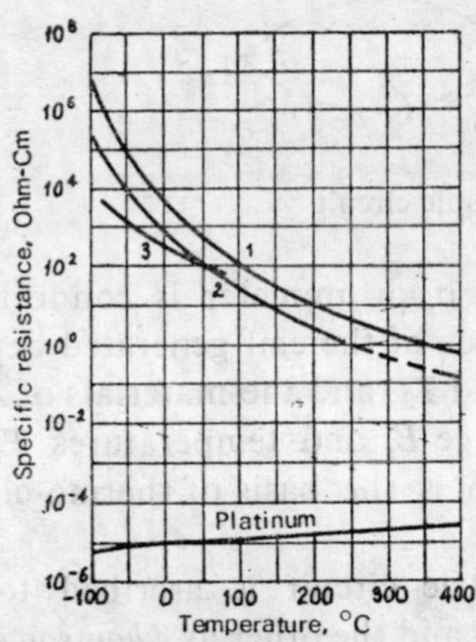

Fig. 8.9 Resistance-temperature characteristics of thermistors

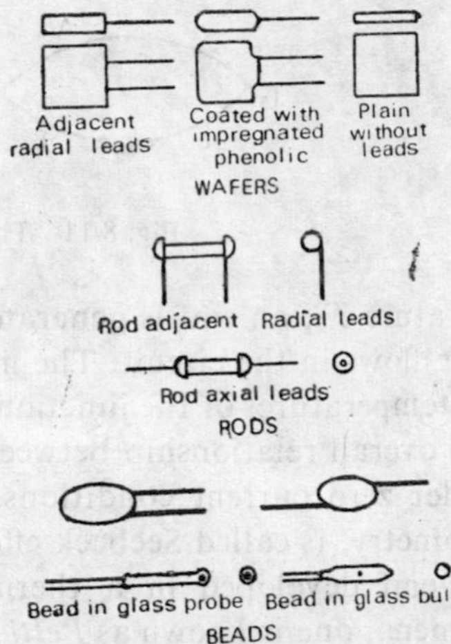

Fig. 8.10 Typical constructions of thermistors

other discs or rods are pigtail-mounted. Beads and small discs may be covered with a thin adherent coat of glass to reduce composition changes of thermistor at high temperature.

The temperature-resistance relationship of the thermistor is given by

$$R = R_0 e^K$$

where $K = \beta(1/T - 1/T_0)$

R = resistance at temperature T°K,

R_0 = resistance at temperature T_0°K, and

β = constant.

The value of β usually lies between 3400/°K and 3900/°K depending on the formulation or grade. The temperature coefficient of resistance is

$$\frac{dR/dT}{R} = -\frac{\beta}{T^2}$$

Assuming β = 4000/°K and T = 298°K, $(dR/dT)/R = -0.045$.

The value of $(dR/dT)/R$ for platinum is 0.0036, indicating that the thermistor is at least 10 times more sensitive than the platinum resistance element.

The application of thermistors to the measurement of temperature follows the usual principles of resistance thermometry. Conventional bridge or other resistance measuring circuits are commonly employed. The high temperature coefficients of thermistors result in their having greater available sensitivity as temperature-sensing elements than the resistance thermometers or common thermocouple.

Thermo-electric thermometry

If two wires of different metals A and B are connected in the circuit as shown in Fig. 8.11, with one junction at temperature T_1 and other at

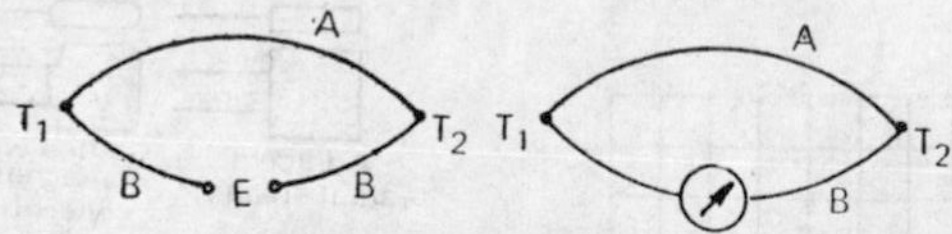

Fig. 8.11 Thermocouple circuit

temperature T_2, an emf is generated, and if an ammeter is connected, a current flows in the circuit. The magnitude of the emf generated depends on the temperatures of the junctions T_1 and T_2, and the materials of A and B. The overall relationship between voltage E_s and temperatures T_1 and T_2 under zero current conditions, which is the basis of thermo-electric thermometry, is called Seebeck effect.

The emf developed in a thermo-electric circuit is ascribed to two phenomena; one is known as *Peltier effect* and the other as *Thomson effect*. Peltier effect concerns the reversible evolution, or absorption, of heat that usually takes place when an electric current crosses a junction between two dissimilar metals. This effect takes place whether the current is introduced externally or is induced by the thermocouple itself. External heating, or cooling, of the junction results in the reversal of the Peltier effect, i.e., a net electric current will be induced in one direction. Thomson effect concerns the reversible evolution, or absorption of heat, occurring whenever an electric current traverses a single homogeneous conductor, across which a thermal gradient is maintained regardless of external introduction of current or its induction by the thermocouple itself. The magnitude and direction of the Thomson voltage set up in a single conductor depend upon temperature level, temperature difference, and the material. The Thomson voltage alone cannot sustain a current in a single homogeneous conductor forming a closed circuit, since equal and opposite emf's will be set up in the two paths from heated to cooled parts. The emf generated due to Thomson effect is less predominant than that from the Peltier effect.

The total emf acting in the circuit is the result of four emf's; two due to Peltier effect (one at each junction) and two due to Thomson effect. The Peltier emf's are assumed proportional to the temperature difference of the junctions, while the Thomson emf's are proportional to the difference between the squares of the junction temperatures. For the total emf E_s the equation takes the form

$$E_s = C_1(T_1 - T_2) + C_2(T_1^2 - T_2^2)$$

for Cu-constantan couple,

$$C_1 = 37.5\ \mu\text{V}/°\text{K}$$

and

$$C_2 = -0.045\ \mu\text{V}/(°\text{K})^2.$$

The polarity of the emf depends on the particular metal used and by the relationship of the temperatures at the two junctions.

The emf can be measured either by a millivoltmeter or a potentiometer. When a millivoltmeter is used for the measurement of emf, a current flows in the circuit. According to Peltier law, this will result in the cooling of one junction and heating of the other. This will, therefore, lower the temperature of the hot junction. But the magnitude of this current is so low as to effectively change the temperature. However, a potentiometer arrangement is normally preferred. Materials for thermocouples (commercial) are selected such that Thomson effect can be disregarded and the total emf is the sum of Peltier emf's only and thus depends only on the difference of the junction temperatures. If the temperature at one junction (reference junction) is kept constant, the emf generated is used to measure the temperature change.

Laws of Thermocouples

The following laws are very useful as they govern both theory and practice of thermocouples.

(i) LAW OF HOMOGENEOUS CIRCUIT. *A thermo-electric current cannot be sustained in a circuit of a single homogeneous material, however varying in cross section, by the application of heat alone.* The consequence of this law is that two different materials are required for any thermocouple circuit.

(ii) LAW OF INTERMEDIATE MATERIALS. *Insertion of an intermediate metal into a thermocouple circuit will not affect the net emf, provided the two junctions introduced by the third metal are at identical temperatures.* This law suggests that a device for measuring thermo-electric emf may be introduced into circuit at any point without affecting the net emf, provided all the junctions which are added to the circuit by introducing the device are all at the same temperature. The proof of this statement is given in Fig. 8.12 (a).

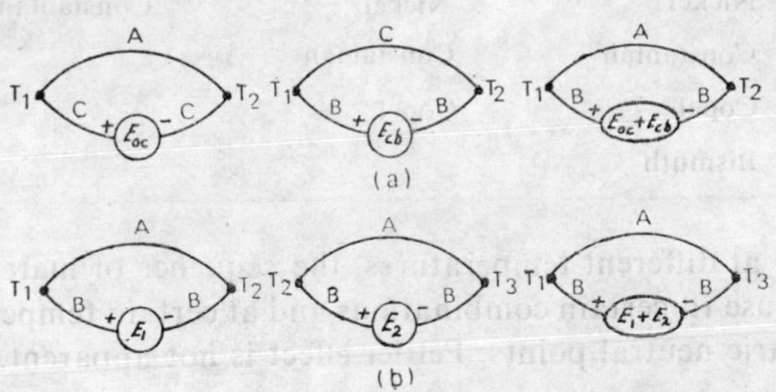

Fig. 8.12 Thermocouple laws

(iii) LAW OF INTERMEDIATE TEMPERATURES. *If a single thermocouple develops a net emf, E, when its junctions are at temperatures T_1 and T_2, and an emf, E′, when its junctions are at T_2 and T_3, it will develop an emf $E_1 = E + E'$, when its junctions are at temperatures T_1* and T_3. The proof of this statement is given in Fig. 8.12 (b). A consequence of this law permits a thermocouple calibrated for a given temperature to be used with any other reference temperature through the use of a suitable correction. Also, the extension wires having the same thermo-electric characteristics as those of the thermocouple wires can be introduced in the circuit without affecting the net emf of the thermocouple.

Thermo-electric series

The various conductors have been tabulated in an order such that at a specified temperature each material in the list is thermo-electrically negative with respect to all above it and positive with respect to all below it.

TABLE 8.2 Thermo-electric series for selected metals and alloys

100°C	500°C	900°C
Antimony	Chromel	Chromel
Chromel	Nichrome	Nichrome
Iron	Copper	Silver
Nichrome	Silver	Gold
Copper	Gold	Iron
Silver	Iron	$Pt_{90}Rh_{10}$
$Pt_{90}Rh_{10}$	$Pt_{90}Rh_{10}$	Platinum
Platinum	Platinum	Cobalt
Palladium	Cobalt	Alumel
Cobalt	Palladium	Nickel
Alumel	Alumel	Palladium
Nickel	Nickel	Constantan
Constantan	Constantan	
Copel	Copel	
Bismuth		

Note that at different temperatures, the sequence of materials changes. This is because in certain combinations and at certain temperatures called thermo-electric neutral points, Peltier effect is not apparent.

Thermocouple materials

Theoretically any two dissimilar conducting materials can be used to form thermocouple. Actually, of course, certain materials and combinations are better than others. These combinations must possess reasonably linear temperature emf relationships, they must develop an emf per degree temperature change that is detectable with standard measuring equipment, they must be able to withstand high temperatures, rapid temperature changes and the effect of corrosive atmosphere. A few preferred materials are:

Copper Cu	Iridium Ir	
Iron Fe	Constantan	(60 Cu + 40 Ni)
Platinum Pt	Chromel	(10 Cr + 90 Ni)
Rhodium Rh	Alumel	(2 Al + 90 Ni + Si + Mn)

TABLE 8.3 Characteristics of some thermocouples

Thermocouple materials	Practical range	Inherent accuracy	Applications
1. Pt—(Pt-10) Rh Pt—(Pt-13) Rh	0°C to 1450°C	0.1°C	Calibration standard where high accuracy at high temperature is required.
2. Cu—constantan	−200°C to 350°C	0.2°C	Commercial oven temperature indication and control.
3. Chromel—alumel	−200°C to 1100°C	0.5°C	Commercial furnace temperature indication and control.
4. Chromel—constantan	−100°C to 1000°C	0.5°C	
5. Iron—constantan	−200°C to 750°C	0.5°C	

The thermocouples of noble metals like Pt—(Pt-Rh) are called precious metal thermocouple and others are base metal thermocouples. Size of the thermocouple wire is of some importance. Usually the higher the temperature to be measured, the thicker should be the wire. As the size is increased, however, the time response of the thermocouple to temperature changes becomes low.

Thermocouples may be prepared by twisting the two wires together and brazing, or preferably welding them as shown in Fig. 8.13. The wires may be separated with the help of insulating beads.

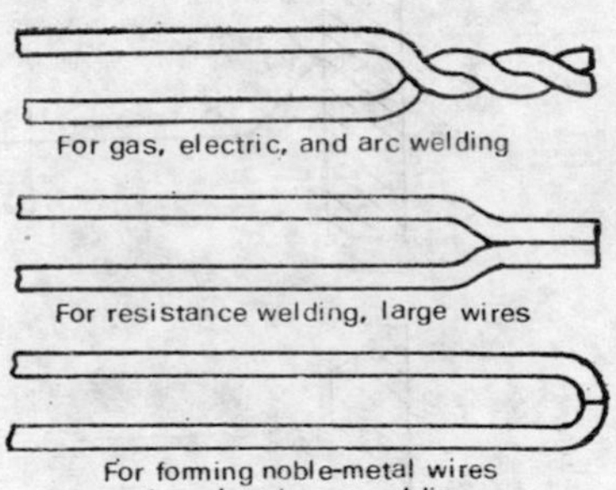

Fig. 1.13 Common forms of thermocouple construction

Measurement of emf

The actual magnitude of the electric voltage developed by thermocouple is very small. Either of the measuring devices is normally employed for determining the thermocouple output—some form of galvanometer, or the voltage balancing potentiometer. The latter is the most commonly used device either as the manually operated or automatically operated. The potentiometer possesses an advantage over the galvanometer in that by basing its operation on buckling emf, no current flows in the thermocouple circuit at balance. Hence, any resistance problems in the leads are largely eliminated. Fig. 8.14 (a) illustrates a simple temperature measuring system employing thermocouple as a sensing element and potentiometer as an indicating instrument. This circuit consists of a measuring junction p and somewhat less obvious junction q at the potentiometer. The temperature of reference junction is read by placing a thermometer near the junction.

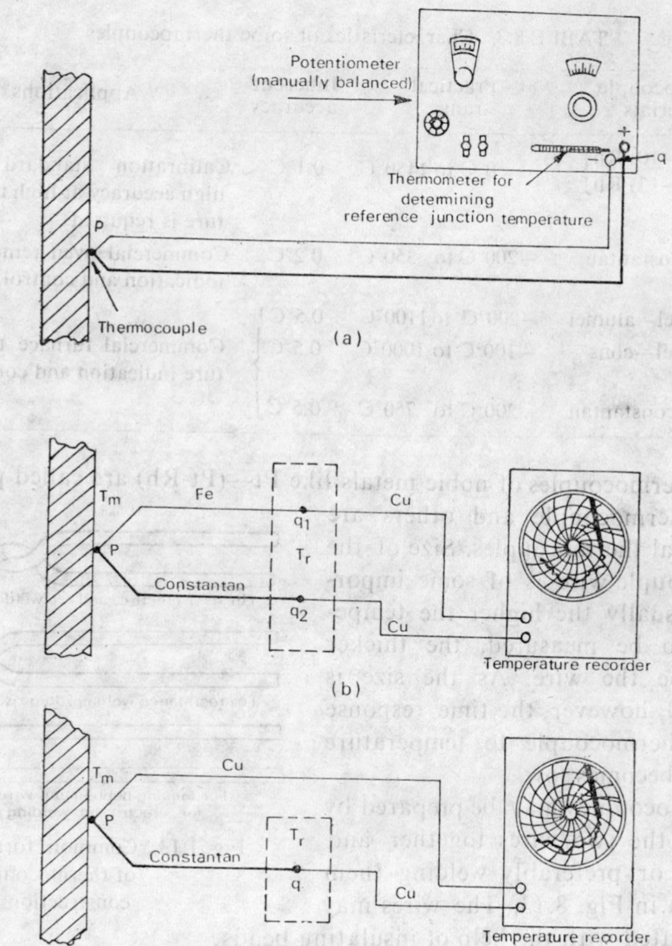

Fig. 8.14 Measurement of EMF

Thermocouple wires are relatively expensive compared to most common materials like copper. It is, therefore, desirable to minimise the use of more expensive material by using extension leads. These arrangements are shown in Fig. 8.14 (b). Of course, a requirement for accuracy is that q_1 and q_2 should be at the same temperature. As indicated above, the reference junction temperature T_r, should be known. The effect of ambient temperature may, therefore, be strong and hence compensations are built in. But in laboratory T_r is kept at accurately controlled conditions. The common arrangement makes use of the ice baths, Fig. 8.14 (c).

Thermocouples may be connected in series to increase the sensitivity as the output emf will be n times that of a single thermocouple provided there are n junctions contributing to the output. This arrangement is called a thermopile.

Thermocouple burn-out feature

Apart from indicating temperature thermocouples are also used for controlling it. As the temperature of measuring junction decreases, its emf decreases. In a controller, the response to a decreasing emf signal is to increase heat input process. In the case of a burned-out thermocouple, no emf is generated creating the danger of maximum heat input to the process taking place.

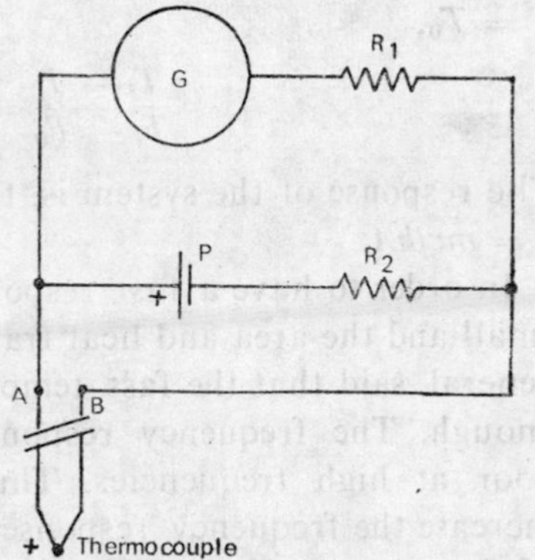

Fig. 8.15 Schematic of thermocouple burnout feature

In most controllers, the methods used are similar to the one explained in the Fig. 8.15. Let the voltage of power source P be E. The resistance of thermocouple including the resistance of the connecting wires to A and B is R_3. The voltage drop across the thermocouple and the lead wires due to current from P is

$$E_1 = \frac{ER_3}{R_2 + R_3}$$

This voltage will cancel the emf generated by the thermocouple. However, this effect can be taken into account in the calibration process itself.

If the thermocouple burns out, the buckling effect from P disappears, and the result is that a current flows through R_1 and galvanometer. The voltage drop caused by this current across the galvanometer and the calibration spool is then $E_2 = ER_1/(R_1 + R_2)$. In practice $E_2 > E_1$ and hence controller will shut down the heat. As a numerical example, consider $E = 1$ volt, $R_1 = 500\ \Omega$, $R_2 = 5000\ \Omega$, $R_3 = 1\ \Omega$, then $E_1 = 0.0002$ volts $= 0.2$ millivolts, equivalent to a temperature change of approximately 4°C, which is compensated. Further $E_2 = 0.091$ volts $= 91$ millivolts.

Dynamics of thermocouple

A thermometer (thermocouple) can be considered as an element with mass m, specific heat c and area A. Let the heat transfer coefficient be h. Insert the thermometer in a temperature field T_2. Then using a very simplified treatment,

$$hA(T_i - T) = mc\frac{dT}{dt}$$

or
$$\frac{dT}{T_i - T} = \frac{hA}{mc}dt$$

On integration,

$$\log(T_i - T) = -\frac{hA}{mc}t + K$$

where K is a constant. Applying the boundary condition that at $t = 0$, $T = T_0$,

$$\frac{T_i - T}{T_i - T_0} = \exp\{-(hA/mc)t\}$$

The response of the system is, therefore, first order with a time constant $\tau = mc/hA$.

In order to have a fast response, the mass and specific heat must be small and the area and heat transfer coefficient should be large. It is, in general, said that the fast temperature changes cannot be followed fast enough. The frequency response depends on the time constant and is poor at high frequencies. Thus a compensation network is used to increase the frequency response of the thermocouple. The disadvantage of the compensation network is that it reduces the thermocouple output. A typical thermocouple compensation network is shown in Fig. 8.16. The thermocouple voltage is represented by E_i and the output voltage by E_0.

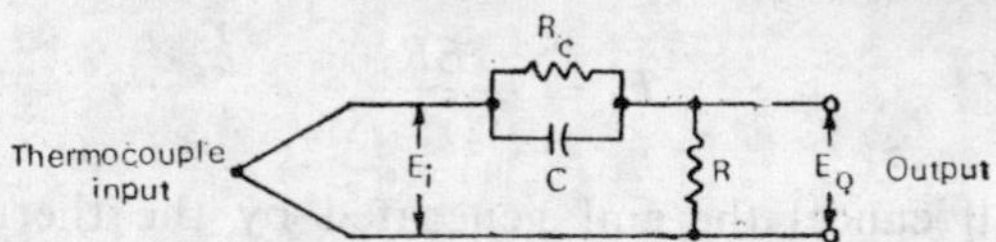

Fig. 8.16 Thermometer compensation network

The transfer function of the system is written as

$$\frac{E_0}{E_i} = \frac{Z_2}{Z_1 + Z_2}$$

where Z_1 is the impedance due to R_c and C, and Z_2 due to R, respectively.

Writing for Z_1 and Z_2 as

$$Z_1 = \frac{R_c \times (1/CS)}{R_c + (1/CS)}$$

where S = Laplace operator,

$= i\omega$ for frequency response

and $Z_2 = R$.

On substitution,

$$\frac{E_0}{E_i} = \alpha\frac{1 + i\omega\tau}{1 + i\omega\alpha\tau}$$

where $\alpha = \dfrac{R}{R + R_c}$ = steady state output,

$\tau = R_c \cdot C$, and

ω = angular frequency of the signal.

The amplitude response is, therefore, given by

$$\left|\frac{E_0}{E_i}\right| = \alpha \sqrt{\frac{1 + \omega^2\tau^2}{1 + \omega^2\alpha^2\tau^2}}.$$

This particular network attenuates smaller frequencies more than high frequencies and has a response to step input which is approximately opposite to that of the thermocouple. High frequency compensation of the network is improved as the value of α is decreased. This brings about a decreased output.

Comparison between resistance thermometer and thermocouple

(i) Resistance thermometers are more accurate than thermocouples. They are usually rated at ± 0.3°C. Thermocouples always offer a slight possibility of inaccuracy due to changes in the reference junction temperature. The errors due to this could be 1.0°C or more.

(ii) Resistance thermometers have greater sensitivity than thermocouples.

(iii) Earlier the response of resistance thermometer element was slower than that of thermocouple. Response of modern resistance elements is about the same as that of a thermocouple.

(iv) Thermocouples require a somewhat more frequent replacement. But thermocouples are cheaper as compared to resistance thermometer elements. Thus final cost of either arrangement is same. The inconvenience caused in replacing the thermocouple might throw the balance in favour of resistance thermometers.

(v) In general, the resistance temperature is used in preference to the thermocouple wherever possible. The principal limitation is temperature.

8.7 Pyrometry

So far temperature-measuring devices kept in contact with the body whose temperature is to be measured have been considered. If the body is very hot, the contact of the thermometer with the body is likely to damage it completely. In other cases a hot body may be far away and hence the question of contact does not arise. In such situations, one relies on instruments which do not require to be placed in contact and the distance between the source and the instrument does not affect the measurement. These instruments are called pyrometers and are based on radiation thermometry. There are two distinct types of pyrometers: total radiation pyrometer, and optical pyrometer.

Total radiation pyrometer, as the name implies, accepts a controlled sample of total radiation, and through determination of the heating effect of the sample, obtain a measure of temperature. Optical pyrometers are again of two types: photon flux, and monochromatic. In one the output from a photosensor indicates the temperature while in the other spectral intensity is compared with a calibrated standard. Before discussing theory and working of pyrometers, a little of the fundamentals of radiation is discussed here.

8.8 Radiation fundamentals

All bodies above absolute zero radiate energy; this radiation depends upon its temperature. The ideal thermal radiator is called a *black body*. Such a body would absorb completely any radiation falling on it. Also, for a given temperature, it emits the maximum amount of thermal radiation possible. The law governing this ideal state is Planck's law which states that

$$W_\lambda = \frac{C_1}{\lambda^5(e^{C_2/\lambda T} - 1)}$$

where W_λ = hemispherical spectral radiant intensity W/cm²-μ,

$C_1 = 37{,}413$ W-μ⁴/cm² (constant),

$C_2 = 14{,}388$ μ-°K (constant),

λ = wavelength in μ, and

T = temperature in °K.

The quantity W_λ is the amount of radiation emitted from a flat surface into a hemisphere, per unit wavelength, at the wavelength λ. Fig. 8.17 illustrates the variation of W_λ and λ for various values of temperature.

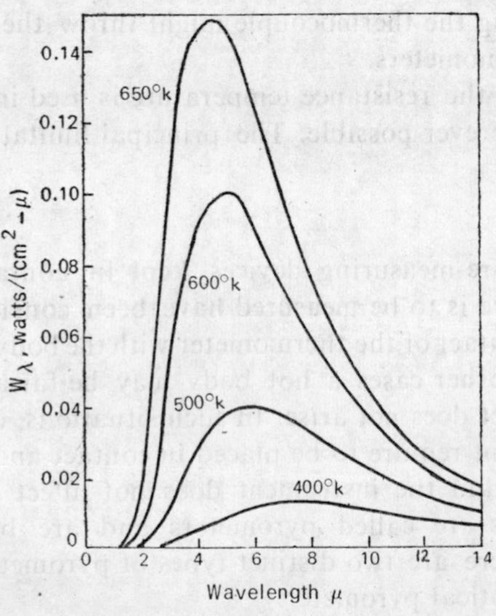

Fig. 8.17 Black body radiation

The W_λ vs. λ curves exhibit peaks at particular wavelength and the peak occurs at high wavelength as the temperature decreases. The maximum of this curve is obtained as

$$\left.\frac{dW_\lambda}{d\lambda}\right|_{\lambda=\lambda_p} = 0$$

This gives $\lambda_p T$ = constant = 2,891 μ-°K. This relation is known as Wien's displacement law.

Further the area under the curve gives the total energy emitted by a black body at a particular temperature and is given by

$$W_t = \int_0^\infty W_\lambda \, d\lambda = \sigma T^4 \quad \text{W/cm}^2$$

This is the total power radiated by a flat surface into a hemisphere. The above relation is called Stefan-Boltzmann law; σ is the Stefan's constant and its value is

$$\sigma = 5.67 \times 10^{-12} \quad \text{W/cm}^2/°\text{K}^4$$

While the concept of black body is a mathematical abstraction, real physical bodies can be constructed to approximate closely to a black body behaviour. The deviations from black body radiation are expressed in terms of emittance of the measured body. Let us call the actual hemispherical spectral radiant intensity of a real body at temperature T as $W_{\lambda a}$ and assume that it can be measured. Then the hemispherical spectral emittance $\epsilon_{\lambda, T}$ is defined as

$$\epsilon_{\lambda, T} = \frac{W_{\lambda a}}{W_\lambda}$$

Thus for $W_{\lambda a}$ one can write

$$W_{\lambda a} = \frac{\epsilon_{\lambda, T} C_1}{\lambda^5 (e^{C_2/\lambda T} - 1)}$$

Similarly, the total power W_{ta} of an actual body is given by:

$$W_{ta} = \int_0^\infty \frac{\epsilon_{\lambda, T} C_1}{\lambda^5 (e^{C_2/\lambda T} - 1)} \, d\lambda$$

and assuming that W_{ta} can be measured, the total hemispherical emittance is

$$\epsilon_{t, T} = \frac{W_{ta}}{W_t}$$

Thus if $\epsilon_{t, T}$ is known, the total power radiated by a real body at temperature T is

$$W_{ta} = \epsilon_{t, T} \sigma T^4$$

One can ascribe a black body temperature T' to any real body by $\sqrt[4]{\epsilon_{t, T}}\, T = T'$, such that

$$W_{ta} = \sigma T'^4.$$

A body whose $\epsilon_{\lambda, T}$ is independent of wavelength at a temperature T, is called a grey body. In this case, $\epsilon_{\lambda, T} = \epsilon_{t, T}$: Since many radiation thermometers operate in a restricted band of wavelengths, the hemispherical band emittance $\epsilon_{t, T}$ has been defined as

$$\epsilon_{b, T} = \frac{\int_{\lambda_a}^{\lambda_b} \{\epsilon_{\lambda, T} \cdot C_1/\lambda^5(e^{C_2/\lambda T} - 1)\} \, d\lambda}{\int_{\lambda_a}^{\lambda_b} \{C_1/\lambda^5(e^{C_2/\lambda T} - 1)\} \, d\lambda}$$

This is seen to be just the ratio of the total powers, of actual and black bodies, within the wavelength intervals λ_a and λ_b when they are at temperature T. If the actual power can be measured directly, $\epsilon_{b, T}$ can be found without knowing $\epsilon_{\lambda, T}$. For grey bodies $\epsilon_{b, T} = \epsilon_{\lambda, T}$.

8.9 Unchopped (d.c.) broad band radiation thermometers

These instruments use blackened thermopile or bolometer as detector and focus the radiation by means of either lenses or mirrors. The lenses have, however, selective transmission and have wavelength dependant focal lengths, unless specifically designed. The use of mirrors rather than of lenses is an attempt to eliminate some of these problems. Fig. 8.18 shows the arrangements used.

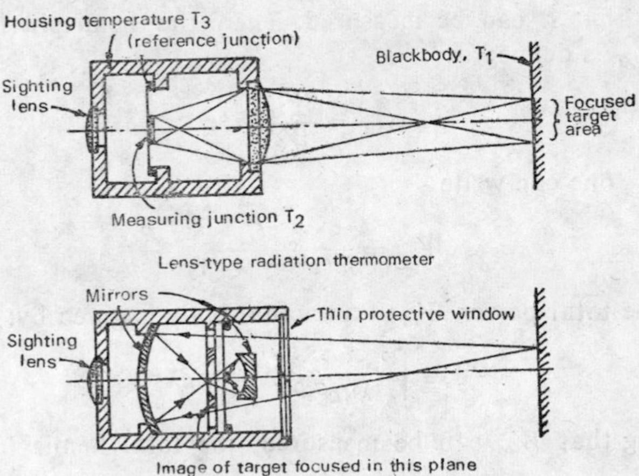

Fig. 8.18 Lens- and mirror-type radiation thermometer.

Basically, for a given source temperature T_1, the incoming radiation heats the measuring junction (when thermocouple is the sensor) until conduction, convection, and radiation losses just balance the heat input. The measuring junction is usually less than 40°C above its surrounding even if the source is incandescent. An over-simplified analysis gives

Heat loss = radiant heat input

Thus

$$K_1(T_2 - T_3) = K_2 T_1^4$$

If the thermocouple voltage is proportional to $T_2 - T_3$, which is usually the case, the voltage output is proportional to T_1^4. For high temperatures, the results obtained are quite good. The temperatures T_2 and T_3 are both influenced by the environmental temperature; thus the compensation is achieved by thermostatically controlling the housing temperature. Calibration of the instrument is generally independent of distance, as long as the target fills the field of view. For smaller distances, focusing is required which may upset the calibration.

The design of a thermopile is shown in Fig. 8.19. All junctions are so arranged as to occupy a very small space, since a focused beam falls over it. Further, the other ends (reference junctions) are fixed on a massive support. The thermopile, very often, is thermostatically controlled to eliminate the effect of ambient temperature.

Fig. 8.19 Thermopile

The thermopile may have 1 or 2 to 20 or 30 junctions. A small number of junctions has less mass and hence a fast response, but low sensitivity. Response of these systems is roughly first order with time constants ranging from 0.1 s to 2 s. The instruments are available to measure temperatures down to −18°C giving a negative output. Theoretically there is no limit to the temperature which can be measured with this instrument. Commercial instruments are available which measure temperature as low as −18°C and as high as 1760°C.

8.10 Chopped (a.c.) broad band radiation thermometer

A number of advantages accrue when the radiation from the target to the detector is periodically chopped at a fixed frequency. Therefore, many infrared systems employ this technique; when sensitivity is needed, amplification is required. High gain a.c. amplifiers are easy to construct than their d.c. counterparts. This is usually the main reason for chopping. Further, unwanted d.c. signal is eliminated by tuning the a.c. amplifier. Additional benefits related to ambient temperature and reference junction compensation may also be obtained.

The effective use of chopping requires a very fast response. Therefore, fast response bolometers or thermistors instead of thermocouple/thermopiles are used as radiation sensors. A mirror focuses the radiation on the detector and a blackened chopper is placed in front of the detector. The chopper periodically cuts off the radiation to the detector. Thus the detector 'sees' radiation alternately from target and the chopper's blackened surface. For high target temperature measurement, sufficient accuracy may be achieved by leaving the chopper at ambient temperature. Higher accuracy particularly at low target temperatures is obtained by thermostatically controlling the chopper temperature. A schematic diagram of one such system is given in Fig. 8.20. A typical instrument uses a square thermistor detector and a field of view of size 1° by 1°. The chopping frequency is 180 cps, leading to an overall time constant of 0.008 s. Temperature range is from ambient temperature to 1300°C and the distance range from 50 cm to infinity, without loss of calibration.

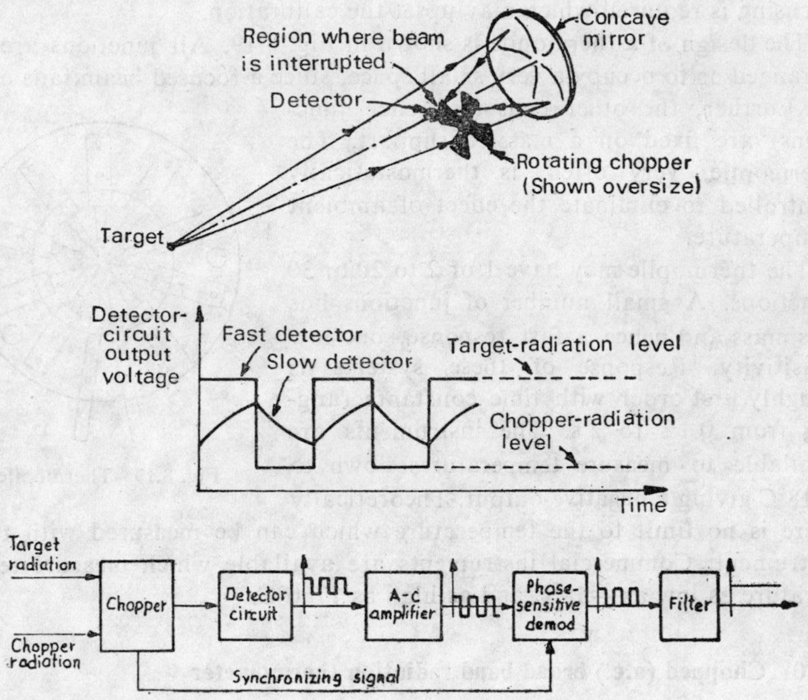

Fig. 8.20 Chopped radiation thermometer.

Accuracy of measurement: The calibration of total radiation pyrometers is done with black body radiation. The output is proportional to T^4. When the pyrometer is employed to measure the temperature of a real body, the knowledge of its emissivity is required. If the emissivity is not accurately known, the measured temperature value will be in error. The magnitude of the error can be obtained as follows:

The output e is given by

$$e = K\epsilon T^4$$

where K is a constant. Differentiating,

$$\left|\frac{dT}{T}\right| = \frac{1}{4}\frac{d\epsilon}{\epsilon}$$

Thus a 10% error in the value of emissivity will result in 2.5% error in the measurement of temperature.

8.11 Optical pyrometers

Chopped (a.c.) selective band (photon flux) radiation thermometers

The use of photon detectors (photo cells, photo transistor, or photo-multiplier tube, etc.) allows faster response and reduces the effect of ambient temperature because it measures the photon flux rather than temperature and this measurement is independent of temperature so long its responsivity with temperature does not vary.

When the radiation is expressed in terms of photon flux rather than Watts, the Planck's formula is modified. The photon flux is given by

$$N_\lambda = \frac{2\pi c}{\lambda^4(e^{C_2/\lambda T} - 1)}$$

where N_λ = hemispherical spectral photon flux (photon/cm²-s-μ)

and c = velocity of light

The maximum of N_λ-λ curve appears at wavelength different from that of radiant intensity. It is given by

$$\lambda_{p,p}T = 3{,}669\mu\text{-}^\circ\text{K}$$

The total photon flux can be obtained by integrating N_λ overall the wave-length, i.e.,

$$N_t = \int_0^\infty N_\lambda \, d\lambda = 1.52 \times 10^{11} T^3 \text{ photon/cm}^2\text{ s}$$

The basic arrangement of the instrument using a photon detector and a chopper, used for the measurement of temperature, is given in Fig. 8.21.

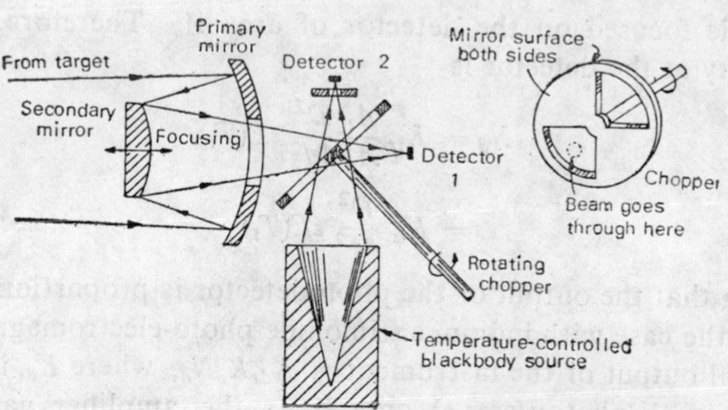

Fig. 8.21 Photon-detector system

A very simplified treatment of the working is given here. Let the distance between the source and the lens/mirror be d_2 and between lens/mirror and detector be d_1. Further the source and detector areas are given by A_t and A_d, respectively. Assuming that the source is focused on the detector and its image fills the detector, then using basic optical laws,

$$\frac{1}{d_1} + \frac{1}{d_2} = \frac{1}{f}$$

and
$$A_t = A_d \frac{d_2^2}{d_1^2}$$

Combining these two relations and assuming $d_2 \gg f$,
$$A_t = A_d \left(\frac{d_2}{f}\right)^2$$

The focal length, therefore, determines the resolution of the instrument, i.e., the minimum target area whose average temperature is determined. The photon flux incident on the lens is given by
$$N_L = N_t \frac{A_t(\pi D^2/4)}{\pi d_2^2}$$

where
$$N_t = \int \epsilon_\lambda N_\lambda \, d\lambda$$
$$= \epsilon \int N_\lambda \, d\lambda \quad \text{for grey bodies}$$
$$= \epsilon E(T_t)$$

and $(\pi D^2/4)/d_2^2$ is the solid angle subtended by the lens aperture on the target. The factor π in the denominator is normalising constant. Assuming lens transmittance of K_{tr}, the number of photons on the other side of the lens are
$$= K_{tr} A_t \frac{D^2}{4d_2^2} \epsilon E(T_t)$$

This flux is focused on the detector of area A_d. Therefore, the photon flux density at the detector is
$$N_d = K_{tr} \frac{A_t}{A_d} \frac{D^2}{4d_2^2} \epsilon E(T_t)$$
$$= K_{tr} \frac{D^2}{4f^2} \epsilon E(T_t)$$

Assuming that the output of the photodetector is proportional to N_d, this is in fact the case with indium antimonide photo-electromagnetic detector, the overall output of the instrument is $K_{dr} K_a N_d$, where K_{dr} is the detector responsivity (V/photon/cm^2 s) and K_a is the amplifier gain. Thus the instrument output voltage is
$$e = \left[K_{dr} K_a K_{tr} \frac{D^2}{4f^2}\right] [\epsilon E(T_t)]$$

Note that the first bracketted term is a constant of the instrument and the second is a function of target temperature and the emissivity. As long as the target is focused, the output is independent of the distance of the target. The variation of $E(T_t)$ with T for a black body target can be found by experimental calibration of the overall instrument. This is necessitated because the instrument lens does not transmit all the wavelengths and due to wavelength dependent k_{tr}. Assuming that the total flux varies as T^3, the output voltage also varies as T^3. Thus for grey bodies

$$e_v = \epsilon K T^3$$

Since the instrument is calibrated against black body, for non-black body value of ϵ must be known to find T. If the value of ϵ is in error, temperature value will be in error. The temperature error is given by

$$\left|\frac{dT}{T}\right| = \frac{1}{3}\frac{d\epsilon}{\epsilon}$$

Thus a 10% error in ϵ will result in a 3.3% error in temperature.

An instrument of this class which uses mirror optics has a temperature range of 38 to 1150°C (up to 4500°C with calibrated aperture), focusing range from 1 m to infinity, field 0.5°, output signal of 10 MV full scale. For the study of rapid transients, chopper may be turned off and the system operates with just the detector and a.c. amplifier. Transients as short as 10 m s can be measured.

Another instrument of this class accepts radiation only in the wavelength band of 4.8 to 5.6 μ. This band avoids the effect of absorption bands of atmospheric water vapour and CO_2 on instrument's response. The range of the instrument is from 38°C to 540°C, time constants 0.2 to 0.5, focusing range from 500 mm to infinity, calibration accuracy of 1% or 11°C whichever is larger, resolution 0.25% of the range or 1.5°C whichever is larger.

Monochromatic optical pyrometer

This pyrometer is the most accurate of all the radiation pyrometers and is used as a calibration standard above the gold point. However, it is limited to temperatures greater than 700°C since it requires a visual brightness match by a human operator. The classical form of the instrument is the disappearing filament-type optical pyrometer. In this an image of the target is superposed on heated tungsten filament. This tungsten lamp, which is very stable, has previously been calibrated so that when the current through the filament is known, the brightness temperature of the filament is also known. Such a calibration is basically obtained by visually comparing the brightness of black body source of known temperature with that of the tungsten lamp. A red filter which passes only a narrow band of wavelengths around 0.65 μ is placed between the observer and tungsten filament. The observer controls the lamp current until filament disappears in the superposed target image. The brightness of the target and that of the lamp are then equal (Fig. 8.22).

Although theoretically there is no upper temperature limit for optical pyrometer, practically, for long term stability, the lamp filament cannot be operated above a certain current or brightness. This limit corresponds to 1350°C. However, much higher temperatures can be measured by inserting absorbing glass filters before the filament. The glass filters

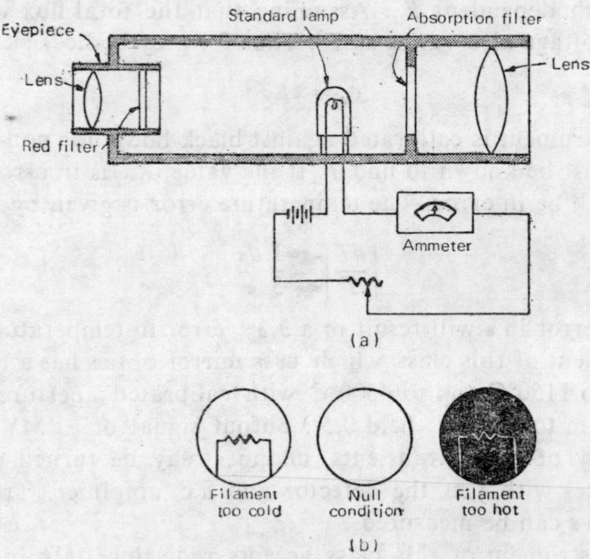

Fig. 8.22 Schematic of an optical pyrometer

attenuate the brightness of the object/target image. A tungsten lamp can be used up to 200 hr before recalibration is required.

The working of optical pyrometer is based on the principle that, at a given wavelength λ, the radiant intensity (brightness) of a black body varies with temperature as

$$W_\lambda = \frac{C_1}{\lambda^5(e^{C_2/\lambda T} - 1)}$$

The lamp is calibrated against a black body, so that the brightness at wavelength λ of the lamp is given by this equation. Therefore, at the moment of equality of brightness of the target and that of the lamp,

$$\frac{C_1}{\lambda^5\{\exp(C_2/\lambda T_L) - 1\}} = \frac{\epsilon_\lambda C_1}{\lambda^5\{\exp(C_2/\lambda_e T) - 1\}}$$

where T_L is the temperature of the lamp, T the temperature of the black body and $\epsilon\lambda_e$ is its emissivity at wavelength λ_e. λ_e is the mean wavelength of the filter $= 0.65\ \mu$. For the temperatures less than 4000°C, the term $\exp(C_2/\lambda T)$ is much greater than 1. Hence, we have

$$\frac{1}{\exp(C_2/\lambda_e T_L)} = \frac{\epsilon_{\lambda_e}}{\exp(C_2/\lambda_e T)}$$

or

$$\frac{1}{T} - \frac{1}{T_L} = \frac{\lambda_e}{C_2}\log \epsilon_{\lambda_e}$$

If the target is a black body ($\epsilon_{\lambda_e} = 1$), there is no error as $T = T_L$. However, if $\epsilon_{\lambda_e} < 1$, but known, this equation allows the calculation of needed correction. The error caused by the inexact value of ϵ_{λ_e} for a particular target is not as great for an optical pyrometer as for an instrument sensitive at wide band of wavelength. The percentage error is given by

$$\left|\frac{dT}{T}\right| = \frac{\lambda_e}{C_2} T \ln \frac{d\epsilon_\lambda}{\epsilon_{\lambda_e}}$$

Thus for a target at 1000°K, a 10% error in ϵ_{λ_e} results only in a 0.45% error in T. The use of red filter aids the operator in making brightness match of the target and the lamp more precisely as colour effects are eliminated. Also target emittance need be known only at one wavelength. If ϵ_{λ_e} is exactly known, the temperature can be measured with optical pyrometers with an error of

3°C at 1,000 °C,

6°C at 2,000 °C,

and 40°C at 4,000 °C.

8.12 Two-colour radiation thermometer

Since errors due to inaccurate values of emittance are a problem in all radiation type temperature measurements, considerable attention has been given to possible schemes for alleviating this difficulty. Although no universal solution has been found, two colour concept has met with some success. The concept requires that W_λ be determined at two wavelengths and their ratio is used to calculate the temperature. For usual conditions of practical applications, the term exp $(C_2/\lambda T)$ is much greater than 1.0 and hence with close approximation one may write as

$$W_1 = \frac{\epsilon_{\lambda_1} C_1}{\lambda_1^5 \exp(C_2/\lambda_1 T)}$$

and

$$W_2 = \frac{\epsilon_{\lambda_2} C_1}{\lambda_1^5 \exp(C_2/\lambda_2 T)}$$

Thus

$$\frac{W_1}{W_2} = \left(\frac{\epsilon_{\lambda_1}}{\epsilon_{\lambda_2}}\right)\left(\frac{\lambda_2}{\lambda_1}\right)^5 \exp\left\{\frac{C_2}{T}\left(\frac{1}{\lambda_2} - \frac{1}{\lambda_1}\right)\right\}$$

for grey bodies $\epsilon_{\lambda_1} = \epsilon_{\lambda_2}$, and hence

$$\frac{W_{\lambda_1}}{W_{\lambda_2}} = \left(\frac{\lambda_2}{\lambda_1}\right)^5 \exp\left\{\frac{C_2}{T}\left(\frac{1}{\lambda_2} - \frac{1}{\lambda_1}\right)\right\}$$

Thus one finds that the ratio $W_{\lambda_1}/W_{\lambda_2}$ is independent of emittance as long as it is numerically the same at λ_1 and λ_2. In one commercial instrument two filters are mounted on a rotating wheel so that the incoming radiation passes alternately through each on its way to photo detector. Instruments covering a range from 1400°C to 4000°C are available.

8.13 Special methods

QUARTZ CRYSTAL THERMOMETER

A novel and highly accurate method of temperature measurement is based on the sensitivity of resonant frequency of a quartz crystal to temperature changes. When the proper angle of cut is used with the crystal, there is a very linear correspondence between the resonant frequency and temperature. Commercial models of the device utilise electronic counters and digital read out for the frequency measurement. For absolute temperature measurement usable sensitivities of 0.001°C are claimed for this device. Since the measurement process relies on frequency measurement, the device is particularly insensitive to noise pick up in the connecting cable, etc.

MICROWAVE TECHNIQUES OF HIGH TEMPERATURE MEASUREMENT

In this instrument again the frequency is measured, the output is digital. The resonant frequency of a cavity is given by

$$f_r = \frac{c}{2}\left[\left(\frac{\lambda}{L}\right)^2 + \left(\frac{Unm}{\pi r}\right)^2\right]^{1/2}$$

where U is the argument for which Bessel function is zero, r is the radius and L the length of the cavity. The cavity expands or contracts with temperature and hence the resonant frequency changes.

Exercises

1. A bimetallic strip of yellow brass ($\alpha_1 = 2.02 \times 10^{-5}$/°C, $E_1 = 0.95 \times 10^6$ N/m²) and Monel 400 ($\alpha_2 = 1.35 \times 10^{-5}$/°C, $E_2 = 1.9 \times 10^6$ N/m²) is bonded at 50°C. The thickness of the yellow brass is 0.35 ± 0.005 mm, and that of the Monel is 0.25 $\pm$ 0.002 mm. If the length of the strip is 125 mm, calculate the deflection sensitivity (deflection per degree centigrade of temperature difference). Estimate the uncertainty in this deflection sensitivity.
2. A bimetallic strip is made by joining two strips, each 150 × 12 × 0.5 mm, one of stainless steel ($\alpha = 1.73 \times 10^{-5}$/°C, $E = 1.93 \times 10^6$ N/m²) and other of iron ($\alpha = 1.08 \times 10^{-5}$/°C, $E = 0.91 \times 10^6$ N/m²). Determine (a) the motion of the free end per unit temperature change (mm/°C), and (b) the force required to hold the strip such that $\delta = 0$ as the temperature increases by a unit amount (kg/°C).
3. The specific volume of mercury is given by the relation $V = V_0(1 + aT + bT^2)$ where T is in °C and

   $$a = 0.1818 \times 10^{-3}$$
   $$b = 0.0078 \times 10^{-6}$$

   A high temperature thermometer is constructed of a Monel 400 tube having an inside diameter of 0.8 ± 0.005 mm. After the inside is evacuated, mercury is filled on the inside such that a column height of 100 ± 0.2 mm is achieved when the thermometer temperature is 260°C. If the tube is to be used for a temperature measurement, calculate the uncertainty at 260°C if the uncertainty in the height measurement is $\pm$ 0.20 mm.
4. A platinum thermoresistive element has the following reported characteristics:

   range −200°C to 500°C,
   sensitivity 0.099 ohm/°C at 0°C,
   resistance 25 ohm at 0°C,
   maximum current 10 mA.

The element is to be used in a bridge as shown in Fig. 8.23. The system is used to measure temperatures in the range 200°C to 300°C. The potentiometer R will be used to adjust the output to zero at 0°C. The fixed resistances are all $\pm 2\%$, and all resistances have a temperature coefficient of 0.00001 Ω/°C. It is desired

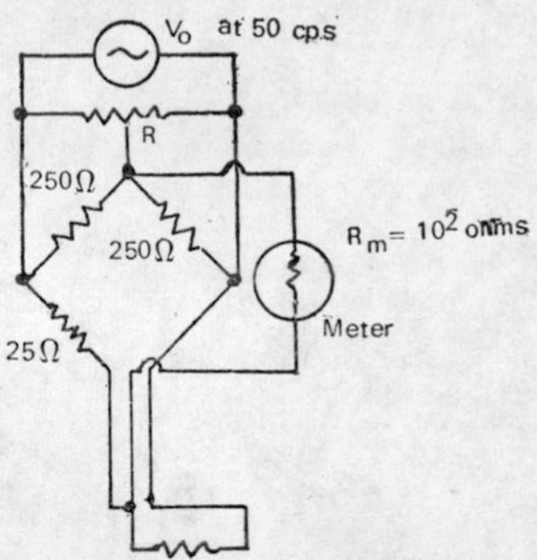

Fig. 8.23

to measure the temperature to a precision of 0.1°C. (i) Determine the supply voltage V_0. (ii) Determine the value of R to assure zero adjustment. (iii) What range and resolution must the meter have? (iv) How large a temperature variation can be tolerated in the bridge circuit?

5. For a certain thermistor, $\beta = 3420$°K, and the resistance at 100°C is known to be 1,040 $\pm$ 3 ohms. The thermistor is used for a temperature measurement, and the resistance is measured as 2,315 $\pm$ 5 ohms. Find the temperature and the uncertainty.

6. An iron-constantan couple is used to measure a temperature of 400°C. The two reference junctions are attached to a copper block to maintain their temperature equal at the room temperature. The meter is connected via copper leads. How much error will be caused in the measurement if the room temperature varies from 20°C to 25°C?

7. For black body radiation, what surface temperature is needed to radiate 100 W per cm^2?

8. Estimate the total power found above the wavelength of 10 μm for black body radiation at $T = 400$°K.

9. When a target of spectral emissivity $\epsilon_\lambda < 1$ is viewed with a monochromatic pyrometer, the measured temperature is somewhat less than the true temperature and will depend on the wavelength at which the measurements are made. The error is given by

$$\frac{T - T_a}{T} = \frac{\lambda T_a}{C_2} \ln \epsilon_\lambda$$

where T = true temperature, and T_a = measured temperature.

For the measurement at 0.665 μm and $T = 1350$°C, calculate $(T - T_a)$ as a function of ϵ_λ.

Bibliography

Aleksandrov, E. B. and Boonch-Bruevich, A. M., "Investigations of surface strains by hologram technique", *Jour. Phys. Tech. Physics* **12**, 258 (1967).

Beckwith, T. G. and N. L. Buck, *Mechanical Measurements*, Addison-Wesley (1961).

Beers, Y., *Introduction to the Theory of Error*, Addison-Wesley (1957).

Behar, M. F., *Handbook of Measurement and Control, Inst. and Automation* **27** (1954).

Benedict, R. P., *Fundamentals of Temperature, Pressure and Flow Measurements*, Wiley (1969).

Boone, P. M., "Holographic Determination of In-plane Deformation", *Opt. Technology* **2**, 94 (1970).

Cerni, R. H., *Instrumentation for Engineering Measurements*, Wiley (1962).

Considine, D. M., *Handbook of Applied Instrumentation*, McGraw-Hill (1964).

Cook, N. H. and E. Rabinowicz, *Physical Measurement and Analysis*, Addison-Wesley (1963).

Doebelin, E. O., *Measurement Systems: Applications and Design*, McGraw-Hill (1966).

Dove, R. C. and P. H. Adams, *Experimental Stress Analysis and Motion Measurement*, Prentice-Hall (1965).

Durelli, A. J. and V. J. Parks, *Moiré Analysis of Strain*, Prentice-Hall (1970).

Ennos, A. E., "Measurements of in-plane surface strain by hologram interferometry", *J. Phys. E. Sci. Inst.* **1**, 73 (1968).

Fourney, M. E., Application of holography to photoelasticity, *Experimental Mechanics*, p. 33 (Jan. 1968).

Haines, K. A. and Hildebrand, "Surface deformation measurement using the wavefront reconstruction technique", *Appl. Opt.* **5**, 595–602 (1966).

Holman, J. P., *Experimental Methods for Engineers*, McGraw-Hill (1966).

Hotzbock, W. G., *Instruments for Measurement and Control*, Rinehold (1962).

Jones, E. B., *Instrument Technology*, Vol. V, Crane-Russak Co.

Kallen, H. P., *Handbook of Instrumentation and Controls*, McGraw-Hill (1961).

Linford, A., *Flow Measurement and Meters*, E. and F. N. Spon, London (1949).

Miesse, C. C. and O. E. Curth, *Product Engineering* (1961).

Index